Introduction to the Physics of Fluids and Solids

James S. Trefil
Clarence J. Robinson Professor of Physics
George Mason University

Dover Publications, Inc., Mineola, New York

Bibliographical Note

This Dover edition, first published in 2010, is an unabridged, slightly cor-
rected republication of the work originally published in 1975 by Pergamon
Press, Inc., New York.

Library of Congress Cataloging-in-Publication Data

Trefil, James S., 1938-
  Introduction to the physics of fluids and solids / James S. Trefill.—
    Dover ed. p. cm.
  Originally published: New York : Pergamon Press, 1975.
  Includes bibliographical references and index.
  ISBN-13: 978-0-486-47437-3
  ISBN-10: 0-486-47437-2
    1. Fluids. 2. Solids. I. Title.

QC145.2.T73 2010
530.4—dc22

                                                          2009025108

International Standard Book Number
ISBN-13: 978-0-486-47437-3
ISBN-10: 0-486-47437-2

Manufactured in the United States by Courier Corporation
47437201
www.doverpublications.com

To my sons
Jim and Stefan

Contents

Preface   **xi**

Chapter 1   Introduction to the Principles of Fluid Mechanics   1
    A. The Convective Derivation   2
    B. The Euler Equation   4
    C. The Equation of Continuity   5
    D. A Simple Example: The Static Star   9
    E. Energy Balance in a Fluid   10
    Summary   11
    Problems   12
    References   15

Chapter 2   Fluids in Astrophysics   16
    SOME APPLICATIONS TO ASTROPHYSICS   16
    A. Equations of Motion   16
    B. The Rotating Sphere   19
    C. Ellipsoids   20
    D. The Earth as a Fluid   27
    E. Jacobi Ellipsoids   27
    F. Rotation of the Galaxy   29
    G. The Rings of Saturn   33
    Summary   36
    Problems   36
    References   38

Chapter 3   The Idea of Stability   **39**
             A. Introduction   39
             B. Stability of the Maclaurin Ellipsoid   43
             Summary   46
             Problems   46
             References   48

Chapter 4   Fluids in Motion   **49**
             A. The Velocity Field   49
             B. The Velocity Potential   56
             C. Stability of Flow   58
             Summary   64
             Problems   64
             References   67

Chapter 5   Waves in Fluids   **68**
             A. Long Waves   68
             B. Surface Waves in Fluids   74
             C. Surface Tension and Capillary Waves   78
             Summary   83
             Problems   83
             References   87

Chapter 6   The Theory of the Tides   **88**
             A. The Tidal Forces   88
             B. Tides at the Equator   89
             C. The Equations of Motion with Rotation   92
             D. Tides at the Surface of the Earth   96
             Summary   102
             Problems   102
             References   104

Chapter 7   Oscillations of Fluid Spheres: Vibrations of the
             Earth and Nuclear Fission   **106**
             A. Free Vibrations of the Earth   106
             B. The Liquid Drop Model of the Nucleus   111
             C. Nuclear Fission   117
             Summary   119
             Problems   119
             References   121

Chapter 8   **Viscosity in Fluids**   **122**
             A. The Idea of Viscosity   122
             B. Viscous Flow Through a Pipe
                (Poisieulle Flow)   127
             C. Viscous Rebound—The Viscosity
                of the Earth   130
             Summary   137
             Problems   137
             References   140

Chapter 9   **The Flow of Viscous Fluids**   **141**
             A. The Reynolds Number   141
             B. Boundary Layers   144
             Summary   151
             Problems   152
             References   154

Chapter 10   **Heat, Thermal Convection, and the Circulation of
             the Atmosphere**   **155**
             A. The Heat Equation and the Bossinesq
                Approximation   155
             B. Stability of a Fluid between Two Plates   159
             C. Convection Cells   167
             D. The General Circulation of the Atmosphere   170
             Summary   178
             Problems   179
             References   183

Chapter 11   **General Properties of Solids—Statics**   **184**
             A. Basic Equations   184
             B. Hooke's Law and the Elastic Constants   188
             C. Bending of Beams and Sheets   189
             D. The Formation of Lacoliths   192
             E. The Formation of Mountain Chains   195
             F. Some Special Cases: Buckling and the Euler
                Theory of Struts   199
             G. Fenno-Scandia Revisited   202
             Summary   204
             Problems   204
             References   208

Chapter 12   **General Properties of Solids—Dynamics**   **209**
                  A. The Strain Tensor   209
                  B. The Stress Tensor   212
                  C. Equation of Motion for Solids   216
                  D. Body Waves in Elastic Media   220
                  E. Surface Waves in Solids   223
                  F. Waves in Surface Layers   227
                  Summary   229
                  Problems   230
                  References   233

Chapter 13   **Applications of Seismology: Structure of the Earth and**
                  Underground Nuclear Explosions   **234**
                  A. Seismic Rays   234
                  B. Underground Nuclear Explosions   240
                  Summary   244
                  Problems   245
                  References   247

Chapter 14   **Applications to Medicine: Flow of the Blood and the**
                  Urinary Drop Spectrometer   **248**
                  A. Introduction   248
                  B. Response of Elastic Arterial Walls to Pressure   252
                  C. Blood Flow in an Artery   256
                  D. The Urinary Drop Spectrometer   264
                  E. Stability of a Capillary Jet   266
                  Problems   270
                  References   274

Appendices   **275**
                  Introduction   275
                  A. Cartesian Tensor Notation   276
                  B. The Gravitational Potential Inside of a
                     Uniform Ellipsoid   279
                  C. The Critical Frequency   283
                  D. Expansion in Orthogonal Polynomials   284
                  E. Solution of Ordinary Differential Equations   288
                  F. The Solution of Partial Differential Equations   292

Index   **301**

Preface

It has become increasingly clear over the past few years that a sizable percentage of the students who leave universities with degrees in physics will not end up doing research in areas normally identified with current research. The increased concern with the environment and with applied research has meant that these students often find themselves working in fields like oceanography or atmospheric physics. In the long-range historical view, this is not strange, since the physicist has traditionally played the role of the generalist in the past. The question about which I have become increasingly concerned is "Are we equipping our students to be the generalists of the future?"

There is a growing body of opinion in the physics community that is coming to the conclusion that this question must be answered in the negative. My own theory about how this state of affairs came about is that we have, to a large extent, stopped teaching physics students about many areas of classical physics. That this should have happened is not surprising, since modern physics research is concerned almost exclusively with quantum systems, such as nuclei, elementary particles, or electrons in a solid. Thus, there is a considerable advantage to the student going into these fields to be introduced to quantum mechanics as soon as possible in his undergraduate career. Unfortunately, this advantage has been gained at the expense of dropping the study of many areas of classical physics from the curriculum. We are now confronted with a situation in which physics graduates may have little or no awareness of the great body of knowledge of fluid mechanics and elasticity which was gained before the beginning of this century.

Ordinarily, this would be unfortunate from a cultural point of view, but would be of little practical importance. The employment situation mentioned above, however, gives the question of education in these fields some urgency, since it is precisely in these areas that most of the applied research will be done. This point was brought home to me most forcefully when I became involved in some interdisciplinary research projects in medicine, and discovered to my chagrin that I did not possess the background necessary to make meaningful contributions in many areas of the research.

After reflecting on these problems, I decided to try to put together a course of lectures which would attempt, in one semester, to allow graduates and advanced undergraduates in physics to learn about these fields. The restriction to a one semester course has the advantage that it does not unduly distort the ordinary course schedules which a student is expected to carry, and the obvious disadvantage associated with trying to cover a lot of material in a short time. My colleagues at the University of Virginia responded to this idea with a great deal of enthusiasm and support, for which I am deeply in their debt, and the course was offered under the title "Topics in Classical Physics." This book is an outgrowth of the course, which has been given for the past three years.

The purpose of this text is twofold. First, an attempt is made to show the student that there is no essential new knowledge which he must master to learn about continuum mechanics. In fact, the basic equations are simply the applications of laws *which he already knows* to new situations. For example, the Euler equation is simply a disguised form of Newton's second law.

Second, it is shown that once these few basic principles are understood, they can be applied to an almost unbelievable number of systems which are seen in nature. Thus, once the laws governing the motion of nonviscous fluids are understood, we can equally well discuss the structure of the galaxy (as in Chapter 2) or nuclear fission (as in Chapter 7).

To emphasize the second point, a large number of examples from many fields of physics have been collected in the text. Partly this is intended to give the flavor of developments in these fields, and partly it is intended to collect, in one convenient location and in one coherent development, problems from as many physics-related fields as possible. Clearly, each reader will have his own taste as to which examples should have been included and which omitted. Space considerations alone would decree that some important areas of physics would have to be left out. Thus, the discussion of stellar structure ignores magnetic and thermal effects, the discussion of blood flow ignores diffusion processes, etc. An instructor using this book as a text can, of course, supply his own examples if he so desires.

The general procedure followed in the development is to introduce a physical principle first, with an emphasis on the nature of the principle and its connection to things already familiar to the student, and then to apply the principle to some interesting system. Sometimes this is done in separate chapters (e.g., Chapter 4 deals with the formalism for dealing with fluids in motion, Chapters 5, 6, and 7 with applications), and sometimes in the same chapters (e.g., Chapter 11 introduces the principles of statics in elastic solids and applies them to geological systems). The mathematical discussion is more or less self-contained, but some appendices on mathematics are included at the end for the sake of completeness.

The completion of a book like this is clearly not the work of a single individual. Many thanks are due both to my colleagues and to the students who acted as subjects for this experiment in physics teaching. Both groups made many valuable suggestions which I have incorporated into the development of the subject.

Special mention should be made of my colleagues in the work on the urinary drop spectrometer (see Chapter 14), Rogers Ritter and Norman Zinner, M.D., who first introduced me to the fascinating field of medical research, and to G. Aiello and P. Lafrance, who have been working and learning with us.

Finally, I would like to thank Mrs. Mary Gutsch for her invaluable assistance in putting the manuscript together, as well as for her refusal to be intimidated by the amount of work involved, and my wife, Jeanne Waples, for her help in the final stages of the organization.

Charlottesville, Virginia   J. S. TREFIL

The Author

James S. Trefil (Ph.D., Stanford University) is an Associate Professor of Physics and Fellow in the Center for Advanced Studies at the University of Virginia. He has published extensively in the area of theoretical high energy physics, and has held visiting positions at several major laboratories in that field. More recently, he has become interested in the applications of physics to medicine, and has contributed to research in the fields of urology, cardiology, and radiobiology.

1

Introduction to the Principles of Fluid Mechanics

Little drops of water
Little grains of sand
Make the mighty ocean
And the pleasant land.

R. L. STEVENSON
A Child's Garden of Verses

Fluids appear everywhere around us in nature. In this section of the book, we shall discuss some of the basic laws which govern the behavior of fluids, and look at the applications of these laws to various physical systems. We shall see that good understandings of the workings of many different types of physical systems can be derived in this way.

Perhaps the most amazing idea that will be developed is that fluid mechanics is not limited in its applications to discussing things like the flow of fluids in laboratories, or the motion of tides on the earth, but that it can successfully be applied to systems as different as the atomic nucleus on the one hand, and the galaxy on the other. Because in dealing with a fluid, we are in reality dealing with a system which has many particles which interact with each other, and because the main utility of fluid mechanics is the ability to develop a formalism which deals solely with a few macroscopic quantities like pressure, ignoring the details of the particle interactions, the techniques of fluid mechanics have often been found useful in making models of systems with complicated structure where interactions (either not known or very difficult to study) take place between the constituents. Thus, the first successful model of the fission of heavy elements was the liquid drop model of the nucleus, which treats the nucleus as a fluid, and thus replaces the problem of calculating the

1

interactions of all of the protons and neutrons with the much simpler problem of calculating the pressures and surface tensions in a fluid. Of course, this treatment gives only a very rough approximation to reality, but it is nonetheless a very useful way of approaching the problem.

A classical fluid is usually defined as a medium which is infinitely divisible. Our modern knowledge of atomic physics tell us, of course, that real fluids are made up of atoms and molecules, and that if we go to small enough scale, the structure of a fluid will not be continuous. Nevertheless, the classical picture will be approximately correct provided that we do not look at the fluid in too fine a detail. This means, for example, when we introduce "infinitesimal" volume elements of the fluid, we do not mean to imply that the volume really tends to zero, but merely that the volume element is very small compared to the overall dimensions of the fluid, but very large compared to the dimensions of the constituent atoms or molecules. So long as we talk about classical macroscopic fluids, there should be no difficulty in making this sort of approximation. Indeed, what is "infinitesimal" is largely a matter of the kind of problem one is working on. It is not at all unusual for cosmologists to consider "infinitesimal" volume elements whose sides are measured in megaparsecs!

A. THE CONVECTIVE DERIVATION

If we are going to describe the motion of fluids, we will have to know how to write Newton's second law for an element of the fluid. This law takes the form

$$\mathbf{F} = \frac{d}{dt}(m\mathbf{v}), \tag{1.A.1}$$

where m is the mass of the element. We are led naturally, then, to consider total time derivatives of quantities which describe the fluid elements. While this may seem straightforward, the fact that the fluid element is in motion makes it somewhat more complicated than it would seem at first glance. To see why this is so, let us consider some quantity f associated with a fluid element (for definiteness, we could think of pressure or entropy or velocity). Then, if the element is at a position x at a time t, at a time $t + \Delta t$ it will be at a new position. (See Fig. 1.1.) Now the definition of a time derivative is

$$\frac{df}{dt} = \lim_{\Delta t \to 0} \frac{f[x(t + \Delta t), \, t + \Delta t] - f[x(t), \, t]}{\Delta t}. \tag{1.A.2}$$

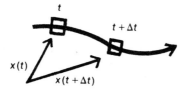

Fig. 1.1. The movement of the volume element.

We see that the fact that in general the function f depends on x, which is itself a function of time, means that some care must be exercised in taking the derivative.

Formally, we can use the chain rule of differentiation to write

$$\frac{\partial f}{\partial t} + \frac{\partial f}{\partial x} dx + \frac{\partial f}{\partial y} dy + \frac{\partial f}{\partial z} dz = \sum_{i=1}^{3} \frac{\partial f}{\partial x_i} dx_i + \frac{\partial f}{\partial t} dt = df, \qquad (1.A.3)$$

where the index i indicates which component of the vector x is being differentiated. (This notation is a trivial example of the method of Cartesian tensors which is discussed in Appendix I.) If we divide through the above by dt, we find

$$\frac{df}{dt} = \sum_i \frac{\partial f}{\partial x_i} \frac{\partial x_i}{\partial t} + \frac{\partial f}{\partial t}. \qquad (1.A.4)$$

But, by definition,

$$\frac{\partial x_i}{\partial t} = v_i,$$

where v_i is the ith component of the velocity of the fluid element. Therefore, the total derivative of the function f with respect to time is just

$$\frac{df}{dt} = \frac{\partial f}{\partial t} + \sum_i v_i \frac{\partial f}{\partial x_i}$$
$$= \frac{\partial f}{\partial t} + \mathbf{v} \cdot \nabla f, \qquad (1.A.5)$$

where we have used the definition of the gradient operator in the latter equality. This total derivative occurs frequently in fluid mechanics, and is given a special name. It is called the convective derivative, and is usually written

$$\frac{D}{Dt} = \frac{\partial}{\partial t} + \mathbf{v} \cdot \nabla. \qquad (1.A.6)$$

To fix this idea firmly in mind, consider the following example: Suppose we have a fluid moving around in a container, where one wall of the container is a movable piston. Now let the function f be the pressure experienced by a particular fluid element. Then the pressure as seen by an observer riding around on the element will vary as a function of time for two reasons—(i) there will be some variation in pressure due to the motion of the piston (this corresponds to the first term in the convective derivative), and (ii) the changes in pressure resulting from the fact that the element moves to different regions of the fluid, where the pressure may be different (e.g., it may be rising to the top of the fluid, where the pressure will be less). This corresponds to the $\mathbf{v} \cdot \nabla$ term in the convective derivative.

B. THE EULER EQUATION

The first fundamental equation of hydrodynamics comes from an application of Newton's second law ($F = ma$) to fluid elements. We know a pressure (defined as a force per unit area) is exerted uniformly everywhere inside a fluid. If we consider a fluid element of length Δx and are A (see Fig. 1.2.), then the net force on the element is

$$F = -[(P + \Delta P)A - PA] = -(\Delta P)A, \qquad (1.B.1)$$

where the minus sign denotes that the force acts in such a way as to cause a flow from regions of higher pressure to regions of lower pressure. If we multiply and divide the right-hand side of the equation by Δx, and note that $\Delta x A = V_0$, where V_0 is the volume, then Newton's law applied to the volume element reads

$$m \frac{d^2 x}{dt^2} = F = -\frac{\partial p}{\partial x} V_0,$$

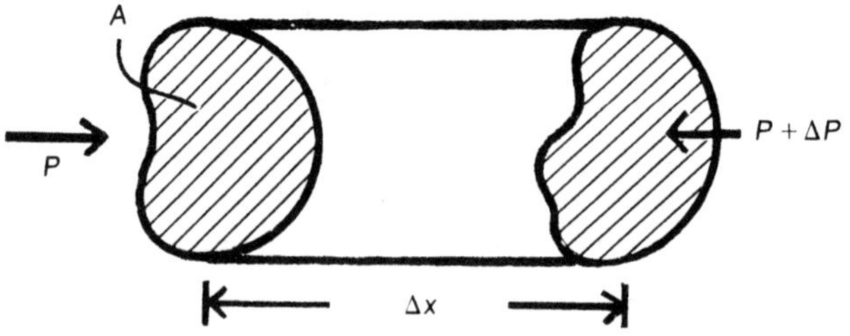

Fig. 1.2.   Forces on a volume element.

or, in terms of the density $\rho = m/V_0$,

$$\rho \frac{Dv_x}{Dt} = -\frac{\partial P}{\partial x},$$

or, in three-dimensional form

$$\rho \frac{D\mathbf{v}}{Dt} = -\nabla P. \tag{1.B.2}$$

The acceleration term of the left-hand side involves a total derivative, so it should really be understood as a convective derivative in the sense of Section 1.A. We should also note that if forces other than pressure (e.g. gravity) were acting on the fluid element, they would appear on the right-hand side of the equation. Thus, the final form of Newton's second law applied to a fluid element is

$$\frac{\partial \mathbf{v}}{\partial t} + (\mathbf{v} \cdot \nabla)\mathbf{v} = -\frac{1}{\rho}\left(\nabla P\right) + \frac{\mathbf{F}_{ext}}{\rho}, \tag{1.B.3}$$

where $\mathbf{F}_{ext}$ is any external force on the fluid element, such as gravity. This first fundamental equation of hydrodynamics is known as the Euler equation.

An alternate form of the equation can be derived if we use the result of Problem 1.1 that

$$\frac{1}{2}\nabla v^2 = \mathbf{v} \times (\nabla \times \mathbf{v}) + (\mathbf{v} \cdot \nabla)\mathbf{v}, \tag{1.B.4}$$

which, when substituted into Eq. (1.B.3) gives

$$\frac{\partial v}{\partial t} + \frac{1}{2}\nabla v^2 - v \times (\nabla \times v) = -\frac{1}{\rho}\nabla P. \tag{1.B.5}$$

If we take the curl of both sides of this equation, and recall that the curl of the gradient vanishes, we get

$$\frac{\partial}{\partial t}(\nabla \times \mathbf{v}) = \nabla \times [\mathbf{v} \times (\nabla \times \mathbf{v})]. \tag{1.B.6}$$

These two alternate forms of the Euler equations will occasionally be useful in dealing with particular physical problems.

C. THE EQUATION OF CONTINUITY

One of the basic precepts of classical physics is that matter can neither be created nor destroyed. The application of this principle to fluid systems will lead us to our second equation of motion, which is usually called the equation of continuity.

Suppose we have a fluid whose density (in general a function of the coordinates and the time) is given by $\rho(x, y, z, t)$ and where the velocity of the fluid elements is given by $v(x, y, z, t)$. Consider a large volume of the fluid V_0 (see Fig. 1.3). The mass of fluid inside the volume is just

$$m_0 = \int_{V_0} \rho \, dV.$$

Now in general fluid will be flowing in and out across the surface S which bounds the volume V_0. To find out what this flow is, consider an element of surface dS. Suppose the fluid next to the surface element has a velocity v_n normal to the surface. Then in a time Δt, all of the fluid in a cylinder of length $v_n \, \Delta t$ and area dS will cross the surface element in time Δt. The total mass of fluid in the cylinder is (see Fig. 1.3) $m = \rho(v_n \, \Delta t) \, dS$ so the total mass outflow per unit time is just

$$\frac{m_{\text{out}}}{\Delta t} = \int_S \rho v_n \, \Delta t \, dS = \int_S \rho \mathbf{v} \cdot d\mathbf{S}, \qquad (1.C.1)$$

where in the second form of the integral, we have adopted the usual convention of writing the surface element as a vector whose length is equal to the area of the element, and whose direction is normal to the surface element.

The conservation of mass which we discussed above requires that the time rate of change of the mass in the volume V_0 be equal to the outflow of mass. This is a requirement that there be no such thing as a source or

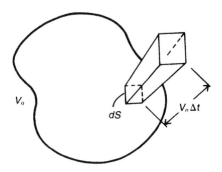

Fig. 1.3. Flow through a closed surface in a fluid.

sink of a classical fluid. Mathematically, we write

$$\frac{\partial}{\partial t} \int_{V_0} \rho \, dV = \int_S \rho v \, d\mathbf{S}, \tag{1.C.2}$$

but Gauss' law says that

$$\int_S \rho v \, d\mathbf{S} = \int_{V_0} \nabla \cdot (\rho \mathbf{v}) \, dV,$$

so that the conservation of mass can be written

$$\int_{V_0} \left[\frac{\partial \rho}{\partial t} + \nabla \cdot (\rho \mathbf{v}) \right] dV = 0. \tag{1.C.3}$$

Since this must be true for any volume inside a fluid, it follows that the integrand itself must vanish, so that we have

$$\frac{\partial \rho}{\partial t} + \nabla \cdot (\rho \mathbf{v}) = 0. \tag{1.C.4}$$

In this form, the requirement of the conservation of mass is called the *equation of continuity*. It will play an extremely important role in our development of fluid mechanics and, together with the Euler equation which we discussed in a previous section, plays the role of one of the basic equations of hydrodynamics.

In our applications of this equation, we shall often deal with *incompressible fluids*. These are fluids for which the density can be considered a constant. In this case, the equation of continuity takes a particularly simple form

$$\nabla \cdot \mathbf{v} = 0 \tag{1.C.5}$$

Suppose we define a fluid current density by

$$\mathbf{j} = \rho \mathbf{v}. \tag{1.C.6}$$

Then the equation of continuity takes the form

$$\frac{\partial \rho}{\partial t} + \nabla \cdot \mathbf{j} = 0. \tag{1.C.7}$$

This is precisely the same equation that one encounters in electromagnetism, where ρ is the charge density and $\mathbf{j}$ is electrical current. The reason for the similarity in the equations, of course, is that just as we postulated that fluid mass can neither be created nor destroyed, in electromagnetism one always postulated that electrical charge is conserved. Our second equation of motion, then, can be thought of as a special case of a more fundamental principle of physics which arises whenever conserved quantities occur in nature.

In the Cartesian tensor notation of Appendix A, the Euler equation can be written

$$\frac{\partial v_i}{\partial t} + v_j \frac{\partial}{\partial x_j} v_i = \frac{1}{\rho} \frac{\partial P}{\partial x_i}. \tag{1.C.8}$$

Since the equation of continuity gives

$$\frac{\partial \rho}{\partial t} = -\frac{\partial (\rho v_k)}{\partial x_k}$$

and

$$\frac{\partial}{\partial t}(\rho v_i) = \rho \frac{\partial v_i}{\partial t} + v_i \frac{\partial \rho}{\partial t}, \tag{1.C.9}$$

this can be rewritten in the form

$$\frac{\partial}{\partial t}(\rho v_i) = -\frac{\partial P}{\partial x_i} - \frac{\partial}{\partial x_k}(\rho v_i v_k),$$

$$= -\frac{\partial}{\partial x_k}(\Pi_{ik}). \tag{1.C.10}$$

where we have defined the two index tensor Π_{ik} by

$$\Pi_{ik} = P\delta_{ik} + \rho v_i v_k. \tag{1.C.11}$$

This tensor is called the *momentum flux tensor*. The reason for this name is quite simple. We know that the momentum of a volume element is just $(\rho V_0)\mathbf{v}$ so that the left-hand side of the above equation is just the time rate of change of the ith component of the momentum of the fluid per unit volume. If we add this up over all of the elements in a volume V_0, we get

$$\frac{\partial}{\partial t} \int_{V_0} \rho v_i \, dV = -\int_{V_0} \frac{\partial \Pi_{ik}}{\partial x_k} \, dV,$$

$$= -\int_S \Pi_{ik} \, dS_k, \tag{1.C.12}$$

where the second equality follows from Gauss' law. Thus, the time rate of change of the momentum in the volume V_0 is the integral of $\Pi_{ik} \, dS_k$ over the surface. Therefore, in analogy to our derivation of the continuity equation, Π_{ik} must be the momentum flux in the ith direction over the kth surface element, and hence represents a net outflow of momentum.

We shall use this momentum tensor form of the Euler equation when we introduce the idea of viscosity later.

D. A SIMPLE EXAMPLE: THE STATIC STAR

The simplest applications of the Euler equation, of course, will be for the case where $v = 0$, the static case. In the next chapter, we will look at many examples of static systems, but for the moment, let us begin by considering a simplified model for a star. We shall see that the two equations which we have derived do not themselves completely specify the system with which we are dealing, but another piece of information will be needed. The extra information is essentially a statement about the kind of fluid of which the system is made.

If we think of a static star, the forces acting on a fluid element will be (i) the pressure and (ii) the gravitational attraction of the rest of the star. This second force is an example of what was called F_{ext} in Eq. (1.B.3). In general, we know that for a gravitational force, we can write

$$F_{ext} = - \rho \nabla \Omega, \tag{1.D.1}$$

where Ω is the gravitational potential. We know that Ω is related to the density of matter by *Poisson's equation*

$$\nabla^2 \Omega = 4 \pi G \rho. \tag{1.D.2}$$

Now the Euler equation in the static case reduces to

$$\frac{1}{\rho} \nabla P = - \nabla \Omega, \tag{1.D.3}$$

which is just the ordinary balance of forces equation from Newtonian mechanics. If we take the divergence of both sides of this equation, we find

$$\nabla \cdot \left(\frac{1}{\rho} \nabla P \right) = - \nabla^2 \Omega = - 4 \pi G \rho. \tag{1.D.4}$$

This is the equation which would have to be satisfied if the star were to be in a state of equilibrium. As it stands, however, it cannot be solved, since it relates two separate quantities—the pressure and the density. What is needed is a relation between these two. This is essentially information about the kind of fluid in the star, since different kinds of fluids will exert different pressure when kept at the same density.

The relation between pressure and density is called an *equation of state*. The reader is probably familiar with one such equation already, the ideal gas law, which says

$$P = R \rho T, \tag{1.D.5}$$

where R is a constant and T is the temperature.

For a star composed of an ideal gas at constant temperature, the equation of equilibrium reduces to

$$\nabla \cdot \left(\frac{1}{\rho} \nabla \rho \right) = -\frac{4\pi G}{RT} \rho. \tag{1.D.6}$$

Specific solutions of this equation are left to the problems.

E. ENERGY BALANCE IN A FLUID

For the sake of completeness, we will discuss the energy associated with fluids, although we shall have few occasions to use this concept in subsequent discussions. Let us consider a fluid in an external field, such as gravity, so that the force is just

$$F = -\rho \nabla \Omega$$

and the Euler equation is

$$\frac{D}{Dt} \mathbf{v} = -\frac{1}{\rho}(\nabla P + \rho \nabla \Omega). \tag{1.E.1}$$

If we take the inner product of the vector $\mathbf{v}$ with this equation, we find, after some manipulation, that

$$\frac{1}{2} \frac{D}{Dt} v^2 = -\frac{1}{\rho} \mathbf{v} \cdot (\nabla P + \rho \nabla \Omega). \tag{1.E.2}$$

If we assume that the potential Ω is independent of the time, so that

$$\frac{\partial \Omega}{\partial t} = 0,$$

then the convective derivative of Ω will be

$$\frac{D}{Dt} \Omega = \mathbf{v} \cdot \nabla \Omega, \tag{1.E.3}$$

so that

$$\frac{D}{Dt} \left(\frac{1}{2} \rho v^2 + \rho \Omega \right) = -\mathbf{v} \cdot \nabla P. \tag{1.E.4}$$

If we note that the total kinetic energy of all of the fluid elements is just

$$T = \int_{v_0} \frac{1}{2} \rho v^2 \, dV$$

and the total potential energy is

$$W = \int_{V_0} \rho \Omega \, dV,$$

then integrating Eq. (1.E.4) over the volume V_0 gives

$$\frac{D}{Dt}(T + W) = -\int_{V_0} \mathbf{v} \cdot \nabla P \, dV, \tag{1.E.5}$$

where the left-hand side represents the total time rate of change of the kinetic plus potential energy of the fluid system. Terms such as this are familiar from other branches of physics. The right-hand side of the equation, however, requires further investigation. If we integrate by parts, we have

$$-\int_{V_0} \mathbf{v} \cdot \nabla P \, dV = -\int_{S} P\mathbf{v} \cdot d\mathbf{S} + \int_{V_0} P(\nabla \cdot \mathbf{v}) \, dV. \tag{1.E.6}$$

The second (volume) integral on the right vanishes for an incompressible fluid. Thus, we are left with the equation

$$\frac{D}{Dt}(T + W) = -\int_{S} P\mathbf{v} \cdot d\mathbf{S}. \tag{1.E.7}$$

The quantity in the integrand has a simple interpretation. $P \, dS$ is just the force acting across the surface element dS (this follows from the definition of the pressure as a force per unit area). This force times the velocity is simply the rate at which the pressure is doing work on the fluid which is crossing the surface element. We see, then, that the above equation is simply the requirement that energy be conserved—that the rate of change of the energy of a fluid system must equal the rate at which work is done across the boundaries.   •

Of course, this is not a new result in the sense that we know that energy must be conserved. Nevertheless, it is comforting to see a familiar law emerge from our formalism.

SUMMARY

In this chapter, we have introduced the basic laws of fluid motion. These laws are seen to follow from some very simple physical principles. These principles are (i) matter can neither be created nor destroyed and (ii) Newton's second law of motion. The principles give rise to the equations of continuity and the Euler equations, respectively.

We saw that these two equations by themselves did not completely define the physics of the simple static star, but that one more piece of information was necessary. This piece of information, in the form of the equation of state, was in reality a specification of the kind of fluid that composed the system. In much of what follows, we will speak of an incompressible fluid—a fluid for which ρ = const. This, too, is an equation of state.

On the basis of these very simple physical principles, a large number of physical problems can be treated, and it is to some of these examples that we now turn.

PROBLEMS

1.1. Using the method of Cartesian tensor notation, show that

$$\epsilon_{ijl}\,\epsilon_{lmn} = \delta_{im}\,\delta_{jn} - \delta_{in}\,\delta_{jm},$$

and prove the following identities

$$\tfrac{1}{2}\nabla v^2 = \mathbf{v} \times (\nabla \times \mathbf{v}) + (\mathbf{v} \cdot \nabla)\mathbf{v},$$

$$\frac{1}{2}\frac{D}{Dt}\,v^2 = \mathbf{v} \cdot \frac{D}{Dt}\,\mathbf{v},$$

$$\nabla \cdot (\mathbf{A} \times \mathbf{B}) = \mathbf{B} \cdot (\nabla \times \mathbf{A}) - \mathbf{A} \cdot (\nabla \times \mathbf{B}).$$

1.2. Show that for a fluid of density ρ at rest in a gravitational field where the acceleration due to gravity at each point in the fluid is $-\mathbf{g}$, that

$$P = g\rho(h - z) + P_0,$$

where z is the vertical coordinate and P_0 is the pressure at a height h, and that the pressure is constant along lines of constant z.

1.3. Show that for an ideal gas at constant temperature, the only solutions to the equation of equilibrium for a star are unphysical (i.e. that they require infinite densities at some point in the star). Are there any values of γ in the *polytropic equation of state* $P = K\rho^\gamma$ for which physical solutions are possible?

1.4. Let us consider vectors and tensors defined in the x-y plane. A rotation in the x-y plane through an angle θ is represented by the matrix

$$R = \begin{pmatrix} \cos\theta & \sin\theta \\ -\sin\theta & \cos\theta \end{pmatrix}.$$

(a) Verify by explicit geometrical construction that the vector

$$\mathbf{v} = a\hat{i} + b\hat{j}$$

transforms according to Eq. (1.A.4).

(b) Verify by explicit calculation and construction that the quantity Π_{ik}, which was defined in Eq. (1.C.11), is indeed a tensor of second rank.

1.5. Consider a fluid where the density varies only with the z-coordinate, so that Poisson's equation becomes

$$\nabla^2 \Omega = \frac{\partial^2 \Omega}{\partial z^2} = -4\pi G\rho$$

and assume further that the fluid is at a constant temperature, so that the equation of state is

$$P = c^2 \rho.$$

Then show that

(a) c is the velocity of sound in the fluid.

(b) The equation for the density is

$$\frac{\partial}{\partial z}\left(\frac{c^2}{\rho}\frac{\partial \rho}{\partial z}\right) = -4\pi G\rho.$$

(c) If the density is taken to be symmetric about the plane $z = 0$

$$\rho = \rho_0 \operatorname{sech}^2\left(\frac{z}{\mathbf{z}}\right),$$

where

$$z = \left(\frac{c^2}{2\pi G\rho_0}\right)^{1/2}.$$

(*Hint*: The change of variables

$$\rho = \rho_0 \Lambda(\xi)$$

and

$$z = \left(\frac{c^2}{2\pi G\rho_0}\right)^{1/2}$$

might prove useful.)

1.6. The force on a moving charge, according to the theory of electrodynamics, is

$$F = q\mathbf{E} + \frac{q}{c}\mathbf{v}\times\mathbf{B},$$

where q is the value of the charge, c is a constant (equal to the speed of light), and E and B are the values of the electrical and magnetic fields which are present.

(a) Consider a fluid which has mass density ρ and charge density σ. Write down the Euler equation for the motion of such a fluid in the case where the fields E and B are fixed by some mechanism external to the fluid.

(b) What is the equation of continuity for ρ? for σ?

1.7. Carry out the energy balance problem of Section 1.E for the fluid described in Problem 1.6. Interpret the new terms which appear in the analogue of Eq. (1.E.7).

1.8. An important thermodynamic property of a material is the entropy per unit volume, s. An adiabatic reaction is defined as a reaction for which the entropy of a system does not change. Show that for an adiabatic reaction,

$$\frac{\partial}{\partial t}(\rho s) + \nabla \cdot (\rho s \mathbf{v}) = 0,$$

where $\rho s \mathbf{v}$ is called the *entropy flux density*.

1.9. One of the most interesting phenomena discovered in the last quarter century is that of the *solar wind*. It was discovered that there are particles around the earth which come from the sun.

(a) Consider a model in which the wind is taken to be the low-density tail of the solar mass distribution. If we assume that the solar particles are static, and that their equation of state is that of an ideal gas, so that

$$P = 2NkT,$$

where N is the number of particles per unit volume, show that the Euler equation requires that

$$0 = \frac{d}{dr}(2nkT) + \frac{GM_s MN}{r^2},$$

where M_s is the mass of the sun and M the mass of a molecule.

(b) It can be shown that the temperature as a function of radius should go roughly as

$$T(r) = T_0 \left(\frac{a}{r}\right)^{1/n+1}.$$

Show that in this case, the number of particles per unit volume, $N(r)$, becomes infinite as $r \to \infty$.

(c) Show that as $r \to \infty$, $p(r)$ approaches a constant which is nonzero. Both parts (b) and (c) show that the solar wind must be a hydrodynamic, as opposed to a hydrostatic phenomenon (as might be guessed from the name).

1.10. Consider the atmosphere as an isothermal gas which has an equation of state given by

$$\rho = a + bP.$$

Determine the pressure as a function of height in such a system, assuming that the earth's surface is flat and does not rotate. Explain where the term "exponential atmosphere" arises.

1.11. Consider a fluid of density ρ moving with velocity $\mathbf{v}$ along the z-axis. Imagine a surface of area dA which is inclined at an angle θ to the z-axis, but which is parallel to the x-axis. Calculate the amount of momentum flow across this surface per unit time by simple mechanics and through the use of the momentum flux tensor defined in Eq. (1.C.11). Show that the results are the same.

1.12. A spherical bathysphere of radius R and mass M descends into the ocean. Assuming that the ocean is made up of incompressible fluid, how far will it sink? Work the same problem for a balloon rising into the air.

1.13. Assuming that water is a fluid of constant density, calculate the force per unit area at the bottom of the Grand Coulee Dam. Why is it thicker at the bottom than at the top?

1.14. Consider a jet of fluid of velocity v and mass M per unit length incident on a plate as shown in the figure. The jet leaves the plate at an angle θ to its original direction, but the plate is arranged in such a way that the magnitude of the fluid velocity does not change. Calculate the force acting on the plate. This is the principle of the turbine.

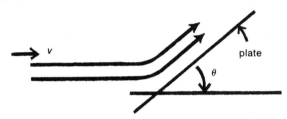

REFERENCES

There are a number of readable books in the field of hydrodynamics, many of which are standard, well-known texts. Some texts of this sort which might be valuable to the reader are

H. Lamb, *Hydrodynamics*, Dover Publications, New York, 1945.
  This book was written in the heyday of classical physics (1879) and revised by the author in 1932. It is an interesting text, mainly because of the large number of examples which are worked out. It is somewhat heavy going for the modern reader, however, because it does not use vector notation.
L. D. Landau and E. M. Lifshitz, *Fluid Mechanics*, Pergamon Press, London, 1959.
  A complete modern exposition of hydrodynamics. The student learning the subject will probably find the mathematical development a little terse, but a large number of topics are covered.
A. S. Ramsey, *A Treatise on Hydrodynamics*, G. Bell and Son, London, 1954.
  A readable book with many examples worked out.
I. Prigogine and R. Herman, *Kinetic Theory of Vehicular Traffic*, American Elsevier, New York, 1971.
  This text applies the ideas of hydrodynamics to traffic flow, and illustrates the remarks made in the Introduction concerning the wide applicability of hydrodynamics.

In addition to the above, many of the texts cited as references in later chapters contain sections dealing with the basic laws of hydrodynamics.

2

Fluids in Astrophysics

There are more things in heaven and earth, Horatio,
Than are dreamt of in your philosophy.

WILLIAM SHAKESPEARE
Hamlet, Act I, Scene V

SOME APPLICATIONS TO ASTROPHYSICS

A. EQUATIONS OF MOTION

On the basis of the basic physical principles which we investigated in the previous chapter, we can now begin to look at some interesting examples of systems in nature. We will begin by considering a uniform fluid which is rotating free from external forces, but where the mutual gravitational attraction of the particles of the fluid for one another is taken into account. This sounds very much like a simple model for an object like a star, and, indeed, the main applications of what we will develop in this chapter have been in the field of astronomy.

We shall begin by investigating the possible equilibrium shapes that a star can have, and then discuss the question of stability. We shall see that it is possible to make definite statements about whether a star could have a certain shape, or whether a star with a certain shape could rotate with a given frequency.

Except where otherwise stated, we shall concern ourselves in this chapter with a fluid which has a constant density. This is an approximation, and, like all approximations, it is good for some systems and not so good for others. It should be pointed out, however, in the spirit of Section 1.D, that this assumption constitutes an equation of state for the system, so that the Euler equation and the equation of continuity will completely define the fluid motion.

Let us consider a mass element in a fluid body (see Fig. 2.1). Let the body be rotating with angular frequency ω about the z-axis. Let r be the vector which describes the position of the element relative to the center of the body, and let $\tilde{\omega}$ be the perpendicular distance from the element to the z-axis (this somewhat clumsy notation is standard for this problem).

Let us now go to a set of axes which are rotating with frequency ω, and are therefore fixed in the body (these are called body axes in classical mechanics). In this system, the body appears to be at rest, so that the velocity of the fluid is everywhere zero. The problem of calculating the motion of the fluid particles is then reduced to the much simpler problem of balancing forces, or *hydrostatics*.

An observer in this system will see the following forces per unit mass acting on a fluid element:

(1) the pressure force, given by $-\dfrac{1}{\rho}\nabla P$,

(2) the gravitational force, given by $-\nabla\Omega$, where Ω is the gravitational potential,

(3) the centrifugal force, given by $\omega^2\tilde{\omega}$.

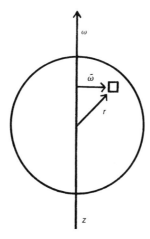

Fig. 2.1. Coordinates for volume elements in a rotating body.

Putting these together, we find for the Euler equation

$$-\omega^2 \tilde{\omega} = -\frac{1}{\rho}\nabla P - \nabla\Omega, \tag{2.A.1}$$

or, in terms of the x-y-z system of coordinates

$$-\omega^2 x = -\frac{1}{\rho}\frac{\partial P}{\partial x} - \frac{\partial \Omega}{\partial x},$$

$$-\omega^2 y = -\frac{1}{\rho}\frac{\partial P}{\partial y} - \frac{\partial \Omega}{\partial y}, \tag{2.A.2}$$

$$0 = -\frac{1}{\rho}\frac{\partial P}{\partial z} - \frac{\partial \Omega}{\partial z}.$$

In all of our applications, we have made the simplifying assumption that the density is not a function of the coordinates. In this case, the first equation can be integrated to give

$$\frac{P}{\rho} + \Omega = \frac{\omega^2 x^2}{2} + f(y, z), \tag{2.A.3}$$

where $f(y, z)$ is an integration "constant" as far as an equation in x is concerned. Differentiating Eq. (2.A.3) with respect to x can convince the reader that there is no way of excluding such an additive function to the solution, just as in ordinary differential equations there is no way of excluding an additive constant from solutions except by applying boundary values.

In a similar way, the remaining Euler equations can be integrated to give

$$\frac{P}{\rho} + \Omega = \frac{\omega^2 y^2}{2} + f(x, z),$$

$$\frac{P}{\rho} + \Omega = f(y, x).$$

The left-hand side of all of these equations is the same quantity, so we can determine something about the arbitrary functions by demanding that the right-hand side of each equation reduce to the same function of the coordinates. In fact, one can readily see that only the choice

$$f(y, z) = \frac{\omega^2 y^2}{2} + C,$$

$$f(x, z) = \frac{\omega^2 x^2}{2} + C,$$

$$f(x, y) = \frac{\omega^2}{2}(x^2 + y^2) + C,$$

where C is a constant will do this. Hence we find for the integrated form of the Euler equation the result

$$\frac{P}{\rho} + \Omega = \frac{\omega^2}{2}(x^2 + y^2) + C. \tag{2.A.4}$$

B. THE ROTATING SPHERE

As a first example of the application of Section 2.A we shall consider a sphere of radius a rotating with angular frequency ω about an axis (see Fig. 2.2).

We begin by calculating Ω, the gravitational potential at a point inside the sphere a distance r from the center. The total mass enclosed within a sphere of radius r is just

$$M(r) = \tfrac{4}{3}\pi r^3 \rho, \tag{2.B.1}$$

so that the potential is just

$$\Omega = \frac{4\pi G \rho r^2}{3}. \tag{2.B.2}$$

Putting this into Eq. (2.A.4), we find

$$2\frac{P}{\rho} = x^2\left(\omega^2 - \frac{8\pi\rho G}{3}\right) + y^2\left(\omega^2 - \frac{8\pi\rho G}{3}\right)$$
$$- \frac{8}{3}\pi\rho G z^2 + 2C. \tag{2.B.3}$$

It then follows that the surfaces of constant pressure will be given by the equation

$$x^2\left(\omega^2 - \frac{8\pi\rho G}{3}\right) + y^2\left(\omega^2 - \frac{8\pi\rho G}{3}\right) - \frac{8\pi\rho G}{3}z^2 = \text{const.} \tag{2.B.4}$$

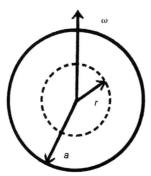

Fig. 2.2.   A rotating sphere.

Now in order to have a stable rotation, it is necessary that the surface of the body be a surface of constant pressure. Otherwise there will be a pressure gradient between two points on the surface and there will not be an equilibrium. The equation for the surface is given by

$$x^2 + y^2 + z^2 = a^2.$$

Clearly, the surface will coincide with a surface of constant pressure only if

$$\omega^2 - \frac{8\pi\rho G}{3} = -\frac{8\pi\rho G}{3}, \qquad (2.B.5)$$

i.e. if

$$\omega = 0.$$

Thus, our investigation of the simplest rotating body—a sphere—shows that it can be in a state of equilibrium only for the trivial case of no rotation. The physical reason for this is quite simple. The centrifugal force tends to throw out material near the equator more than at the poles, so most rotating bodies can be expected to have a somewhat "squashed" appearance. This means that we shall have to turn our attention to more complicated geometries if we want to look at more realistic cases.

Although in the case of a sphere the only solution to our equation is the trivial one of $\omega = 0$, the method we used will be repeated for more complicated geometries, where it will be less easy to follow. To review: to see if there is an equilibrium possible for a rotating fluid, we must

(1) Calculate the gravitational potential inside the fluid.
(2) Insert this potential into Eq. (2.A.4) to determine the surfaces of constant pressure.
(3) Ascertain whether one of these surfaces could coincide with the actual surface of the body.

If the answer to the last step is yes, then an equilibrium is possible—i.e. the body can be rotated without changing its shape.

C. ELLIPSOIDS

The simplest possible equilibrium shape for a rotating gravitating fluid once the sphere has been eliminated is that of an ellipsoid. Consider such a body rotating with frequency ω about its z-axis, which we take to lie along one of the major axes of the ellipse.

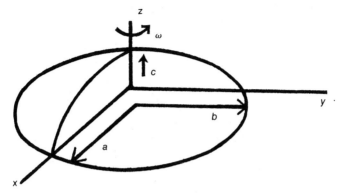

Fig. 2.3. The rotating ellipsoid.

In Appendix B, we show that the gravitational potential analogous to Eq. (2.B.2) for the sphere is just

$$\Omega = \pi \rho \, abc \, G \int_0^\infty \left(\frac{x^2}{a^2 + \lambda} + \frac{y^2}{b^2 + \lambda} + \frac{z^2}{c^3 + \lambda} - 1 \right) \frac{d\lambda}{\Delta}, \qquad (2.C.1)$$

where

$$\Delta = [(a^2 + \lambda)(b^2 + \lambda)(c^2 + \lambda)]^{1/2}, \qquad (2.C.2)$$

which for the sake of convenience we can write

$$\Omega = \pi \rho G (\alpha_0 x^2 + \beta_0 y^2 + \gamma_0 z^2 - \chi_0), \qquad (2.C.3)$$

where

$$\alpha_0 = abc \int_0^\infty \frac{d\lambda}{\Delta(a^2 + \lambda)}, \qquad (2.C.4)$$

and where β_0, γ_0, and χ_0 are similarly defined.

We can now proceed as we did in the case of the sphere, putting the above expression for the potential energy into Eq. (2.A.4), the integrated Euler equations, and demanding that a surface of constant pressure coincide with the surface of the ellipse, which in this case is given by the expression

$$\frac{x^2}{a^2} + \frac{y^2}{b^2} + \frac{z^2}{c^2} = 1.$$

In such a procedure, the information about the shape of the ellipsoid is contained in the constants α_0, β_0, γ_0, and χ_0.

The integrated Euler equation, with the potential for the ellipsoid, becomes

$$\left(\alpha_0 - \frac{\omega^2}{2\pi\rho G}\right)x^2 + \left(\beta_0 - \frac{\omega^2}{2\pi\rho G}\right)y^2 + \gamma_0 z^2 = \frac{1}{\pi\rho G}\left(\frac{P}{\rho} + C\right) + \chi_0. \quad (2.\text{C}.5)$$

The surfaces of constant pressure can be obtained from this by setting the right-hand side of Eq. (2.C.5) equal to a constant. In order for one of these surfaces to coincide with the surface of the ellipsoid, it is necessary that (up to a common multiplicative constant),

$$\frac{1}{a^2} = \alpha_0 - \frac{\omega^2}{2\pi G\rho},$$

$$\frac{1}{b^2} = \beta_0 - \frac{\omega^2}{2\pi G\rho}, \quad\quad (2.\text{C}.6)$$

$$\frac{1}{c^2} = \gamma_0.$$

A case of particular simplicity is that of the ellipsoid of revolution, where we have

$$a = b = \frac{\sqrt{\xi^2 + 1}}{\xi}c, \quad 0 \le \xi \le \infty. \quad (2.\text{C}.7)$$

This corresponds to a body in which the cross section perpendicular to the axis of rotation are circles, and represents the next step in geometrical complication after the sphere. It is called the Maclaurin ellipsoid.

It should be noted that we are already anticipating a result which we shall derive later when we write the relation between a and c as we do in Eq. (2.C.7) because no matter what value of ξ we pick, c will always be less than or equal to a and b. Thus we are considering only oblate spheroids. The prolate spheroid is left to Problem 2.3 at the end of the chapter.

We can now write down the structure constants directly

$$\alpha_0 = \beta_0 = b^2 c \int_0^\infty \frac{d\lambda}{(b^2 + \lambda)^2 \sqrt{c^2 + \lambda}},$$

$$= (\xi^2 + 1)\xi \text{ arc cot } \xi - \xi^2, \quad\quad (2.\text{C}.8)$$

where we have used the change of variables

$$c^2 + \lambda = (a^2 - c^2)a^2,$$

to carry out the integrals. Similarly,

$$\gamma_0 = 2(\xi^2 + 1)(1 - \xi \text{ arc cot } \xi). \tag{2.C.9}$$

The structure constant χ_0 could be computed as well, but since Eq. (2.A.4) contains an arbitrary constant anyway, we can simply incorporate χ_0 into it.

For this simplified geometry, the condition that the surface of the ellipsoid corresponds to a surface of constant pressure reduces to

$$\left(\alpha_0 - \frac{\omega^2}{2\pi_p G}\right)a^2 = \gamma_0 c^2, \tag{2.C.10}$$

which reduces to

$$\frac{\omega^2}{2\pi\rho G} = \xi \cot^{-1} \xi(3\xi^2 + 1) - 3\xi^2, \tag{2.C.11}$$

when the values of α_0 and γ_0 computed earlier are substituted.

There are two important points which can be made about this equilibrium condition. First, we see that the question of whether or not equilibrium can be established depends only on ξ, which is related to a *ratio* of lengths of major and minor axes of the ellipse. Thus, the size of the ellipse does not matter at all provided that the proportions of the axes are such that Eq. (2.C.11) can be satisfied. Thus, a planet or a galaxy with a given ξ (i.e. a given ratio between major and minor axes) will have the same ratio of frequency of rotation to $2\pi G\rho$ at equilibrium (but since $2\pi G\rho$ depends on the density, they need not have the same frequency of rotation).

To find out whether such a solution exists (i.e. whether an ellipsoid in uniform rotation can be in equilbrium), we can look at a graph of the right- and left-hand sides of the equation as a function of ξ. If the line which represents the left-hand side intersects the curve which represents the right-hand side, then Eq. (2.C.11) will have a solution, and the body will be in a state of equilibrium for that value of ξ.

The shape of the right-hand side can be guessed without actually calculating it by noting that as $\xi \to \infty$,

$$\cot^{-1} \xi \to \frac{1}{\xi} \frac{3\xi^2 - 1}{3\xi^2}$$

so that the right-hand side approaches zero from the positive side. Similarly, as $\xi \to 0$, $\cot^{-1} \xi \to \pi/2$ so that the right-hand side goes to zero as $(\pi/2)\xi$. This means that the right-hand side starts from zero, goes positive, and returns to zero, so that there must be a maximum somewhere in between.

The situation is sketched in Fig. 2.4. In general, the left-hand side need not depend on ξ at all (although for most cases of physical interest, it will—see below), so it will appear on the figure as a straight line. There are several distinct cases. In the case corresponding to the line labeled "1", it is possible for the right- and left-hand sides of Eq. (2.C.11) to be equal, and hence for a solution to exist for which an ellipsoid can rotate in equilibrium. For the line labeled "3", this is not the case, and no solution to our problem will exist. Thus, if $\omega^2/2\pi\rho G$ is large enough, it will be impossible for the ellipsoid to rotate in equilibrium. The case separating these two regimes is the line labeled "2", where $\omega^2/2\pi\rho G$ is just equal to the maximum value of the right-hand side of Eq. (2.C.11).

By explicit calculations, it turns out that the value of the right-hand side at its maximum is 0.224, so that the critical case occurs when

$$\frac{\omega}{\sqrt{2\pi\rho G}} = \sqrt{0.224} \approx \frac{1}{2}.$$

In other words, the maximum frequency at which a Maclaurin ellipsoid can rotate is of the order of $\sqrt{2\pi\rho G}$. This is a special case of a more general result which we prove in Appendix C, which says that it is impossible for any body to be in equilibrium if it is rotating faster than a critical frequency ω_c, where ω_c is defined by

$$\omega_c^2 = 2\pi\rho G. \tag{2.C.12}$$

Physically, we can think of this result in the following way: When a mass is rotating slowly, it is possible for the gravitational attraction to overcome the centrifugal force and hold the fluid together. As ω is increased, however, the centrifugal force will become too great, and the fluid will fly apart.

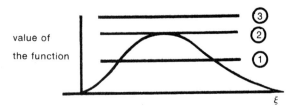

value of
the function

Fig. 2.4.   Plot of the right-hand side of Eq. (2.C.11) as a function of ξ.

This result, while it is valid in the general case, does not shed much light on the problem of classical stellar structure. To understand why, we need to realize that when we discuss a mass of fluid rotating in a vacuum, there are two important quantities which must be conserved. These are the mass and the angular momentum. Since we are dealing with an incompressible fluid, the conservation of mass requires that the volume be fixed as well.

The volume of an ellipsoid of revolution is just

$$V = \frac{4}{3} \pi a^2 c = \frac{4}{3} \pi a^3 \frac{\xi}{\sqrt{\xi^2 + 1}}, \tag{2.C.13}$$

while the moment of inertia about the z-axis is

$$I = \frac{2}{5} M a^2 = \frac{2}{5} M \left[\frac{3}{4\pi} V \frac{\sqrt{\xi^2 + 1}}{\xi} \right]^{2/3}, \tag{2.C.14}$$

so that the angular momentum is

$$L = I\omega = \frac{2}{5} M \left[\frac{3V}{4\pi} \frac{\sqrt{\xi^2 + 1}}{\xi} \right]^{2/3} \omega, \tag{2.C.15}$$

where we have written everything in terms of the conserved quantities M, V, and L and the parameter ξ.

Solving Eq. (2.C.15) for ω and substituting into Eq. (2.C.11), we find

$$\frac{\omega^2}{\omega_c^2} = \frac{25L^2}{4M^2\omega_c^2 \left(\frac{3V}{4\pi}\right)^{4/3}} \left(\frac{\xi}{\sqrt{\xi^2 + 1}}\right)^{4/3}, \tag{2.C.16}$$

$$= \xi \cot^{-1} \xi (3\xi^2 + 1) - 3\xi^2,$$

which can be written

$$\frac{L^2}{L_c^2} = \left(\frac{\xi^2 + 1}{\xi^2}\right)^{2/3} [\xi \cot^{-1} \xi (3\xi^2 + 1) - 3\xi^2], \tag{2.C.17}$$

where we have defined

$$L_c^2 = \frac{4M^2}{25} \left(\frac{3V}{4\pi}\right)^{4/3} \omega_c^2. \tag{2.C.18}$$

The point of this discussion is that whereas in general the frequency of rotation and the parameter ξ can be regarded as independent, when we require that mass and angular momentum be conserved, this is no longer the case, and ω becomes a function of ξ, as in Eq. (2.C.16). This is true whether the equilibrium equation (2.C.11) is satisfied or not.

We can proceed as before, graphing the right- and left-hand sides of Eq. (2.C.16), as shown in Fig. 2.5 where the straight line represents the quantity L^2/L_c^2 (which is now truly independent of ξ, since it depends only on the initial conditions). The point of intersection is the solution which we seek, and represents the configuration at which a given mass ellipsoid with a fixed angular momentum will rotate in equilibrium. We see that for each L, there is one and only one equilibrium configuration for the ellipsoid.

The result in the figure is physically reasonable, since as L is increased, $\xi \to 0$. From Eq. (2.C.7), $\xi \to 0$ corresponds to a flattened out "pancake," so that this agrees with our intuition, which tells us that as we spin a body faster and faster, it will tend to flatten out. Similarly, as L is decreased, $\xi \to \infty$, which corresponds to the ellipsoid approaching a sphere.

Of course, it must be kept firmly in mind that although it appears that there will be a solution to Eq. (2.C.16) for any L, the constraint that ω must be less than ω_c continues to restrict the possible values of L which may be achieved for a given mass.

For the case of the earth, which has mean density 5.52 g/cm^3, this critical frequency is

$$\omega_c = 1.5 \times 10^{-3} \text{ sec}^{-1},$$

which corresponds to a period of

$$T_c = \frac{2\pi}{\omega_c} = 4.25 \times 10^3 \text{ sec} = 1.8 \text{ hr},$$

while for the sun, which has mean density 1.41 g/cm^3, it is

$$\omega_c = 7.7 \times 10^{-4} \text{ sec}^{-1},$$

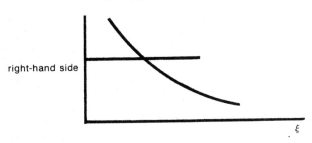

right-hand side

ξ

Fig. 2.5. Plot of the right- and left-hand sides of Eq. (2.C.16).

which is a period of

$$T_c = 8.15 \times 10^4 \text{ sec} = 22.5 \text{ hr.}$$

Thus, both of these bodies rotate at frequencies well below the critical frequency given above.

D. THE EARTH AS A FLUID

In later applications, we shall often wish to treat the earth itself as a fluid mass. Suppose we want to know how realistic such an approximation could be. One measure of such an approximation would be to calculate its rotational frequency from Eq. (2.C.11), and to compare it with the actual frequency of rotation. For the earth, we have

$$a = b = 6.378 \times 10^6 \text{ m}$$

and

$$c = 6.357 \times 10^6 \text{ m,}$$

so that

$$\xi = 12.16,$$

which gives

$$\left(\frac{\omega}{\omega_c}\right)_{\text{predicted}} = 0.059. \qquad (2.\text{D}.1)$$

We can compare this to the observed frequency (taking ω_c from Eq. (2.C.12))

$$\left(\frac{\omega}{\omega_c}\right)_{\text{observed}} = 0.048. \qquad (2.\text{D}.2)$$

These two agree to about 20%, so that if we can be satisfied with that sort of accuracy, we can indeed treat the earth as a fluid mass (even though we know it to be solid). We shall use this result later when we calculate the free vibrations of the earth.

E. JACOBI ELLIPSOIDS

An ellipsoid of revolution in which all three axes are not equal is called a *Jacobi ellipsoid.* For such a configuration, the equilibrium conditions can be cast in the form:

$$\left(\alpha_0 - \frac{\omega^2}{\omega_c^2}\right) a^2 = \gamma_0 c^2, \tag{2.E.1}$$

$$\left(\beta_0 - \frac{\omega^2}{\omega_c^2}\right) b^2 = \gamma_0 c^2. \tag{2.E.2}$$

We could at this point proceed just as we did in the case of the Maclaurin ellipsoid, but recalling the result that for each equilibrium configuration there is just one frequency of rotation which will just balance the forces at the surface, we subtract the above equations to get

$$\frac{\omega^2}{\omega_c^2} = \frac{\alpha_0 a^2 - \beta_0 b^2}{a^2 - b^2}. \tag{2.E.3}$$

Similarly, multiplying Eq. (2.E.1) by b^2 and Eq. (2.E.2) by a^2 and then subtracting gives

$$(\alpha_0 - \beta_0) a^2 b^2 + \gamma_0 c^2 (a^2 - b^2) = 0. \tag{2.E.4}$$

The second of these equations is independent of the frequency. Thus, if we can find a set of values for a, b, and c which satisfy it, we will have the equilibrium configuration. We can then put these values into Eq. (2.E.3) and evaluate the frequency which corresponds to this configuration.

Putting the integral forms for the structure constants in Eq. (2.E.4), we find

$$(a^2 - b^2) \int_0^\infty \frac{d\lambda}{\Delta} \left[\frac{a^2 b^2}{(a^2 + \lambda)(b^2 + \lambda)} - \frac{c^2}{c^2 + \lambda} \right] = 0. \tag{2.E.5}$$

If $a = b$ (the case for the Maclaurin ellipsoid), then this condition is automatically satisfied, and we have found the equilibrium conditions from Section 2.C. If $a \neq b$, however, we can ask the question of whether it is ever possible to satisfy the condition in Eq. (2.E.5).

Instead of solving the problem explicitly, we will show that a solution must exist. To see this, we will consider the value of the integral for two different cases.

Case (i) $c = 0$.

In this case, the second term in the integrand vanishes, and, since λ is always positive, the integral must be positive as well.

Case (ii) $c^2 = \dfrac{a^2 b^2}{a^2 + b^2}$.

In this case, the first term in the integrand becomes $a^2b^2/(a^2b^2 + (a^2+b^2)\lambda + \lambda^2)$ while the second becomes $a^2b^2/(a^2b^2+(a^2+b^2)\lambda)$. Clearly, the second will always be greater than the first, so the integral in this case must be negative.

Thus, we have a situation in which the integral proceeds from a positive value at $c = 0$ to a negative one at $c = a^2b^2/(a^2+b^2)$. At some intermediate point, it must pass through zero, and the values of a, b, and c at this point will give the equilibrium values. For some numerical results, the reader is referred to Lamb (Chapter XII). From these values, the equilibrium rotational frequency can be calculated using Eq. (2.E.3).

One further point should be made. We can divide Eq. (2.E.5) by $(a^2b^2c^2)$ and obtain a form of the equilibrium condition which depends only on the ratios b/a and c/a. This is the scaling result which we saw earlier for the Maclaurin ellipsoid. The equilibrium depends only on relative sizes, and not on the actual magnitude of the dimensions of the rotating body.

F. ROTATION OF THE GALAXY

An interesting application of what has been done so far is to look at the gross structure of the galaxy. One problem of some current interest centers around the galactic rotation curves. These curves are essentially a plot of the velocity of a particle in the galaxy as a function of its distance from the center of rotation. There are several different kinds of rotation curves that one can imagine:

(i) "Solid body" rotation, in which every particle in the galaxy has the same angular frequency as the galaxy as a whole, so that $v(r) \propto r$.

(ii) "Constant velocity" rotation, in which every particle in the galaxy has the same speed (and hence different angular frequencies). In this case, $v(r) = v_0$.

(iii) "Keplerian" rotation, in which particles far from the center see a gravitational force $= Gm/r^2$ which just balances the centrifugal force, and gives $v(r) \propto 1/\sqrt{r}$.

In fact, all three types of rotation are seen in nature. A "typical" rotation curve (such as that for our own galaxy) is shown in Fig. 2.6. We see that at very large distances (where the particles see the rest of the galaxy as a point) we get the expected Keplerian revolution, while for some region of r (which varies from one galaxy to the next) there is constant velocity rotation. At very small r, the rotation becomes solid

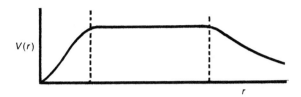

Fig. 2.6. A typical galactic rotation curve.

body. It should be noted that there are galaxies in nature which are predominantly solid body as opposed to the one shown above, which is predominantly constant velocity.

Now the galaxy is obviously a body which is rotating freely under its own gravitational attraction, so that the methods we have developed for treating such bodies are appropriate here. However, we shall see that the main information which can be gained from studying galactic rotation curves has to do with the distribution of matter in a galaxy, so we will want to drop, for the moment, the requirement that the density of the fluid be constant.

The general structure of our galaxy is pictured in Fig. 2.7 (all distances in light years). Most of the mass is concentrated in a central core, but the galaxy is much wider than it is high. This leads us to suppose that we can replace the actual problem of calculating the surface conditions for the rather complicated geometry of the real galaxy by the much simpler problem of calculating for a two-dimensional disk rotating about an axis perpendicular to the plane of the disk.

We can state this supposition with somewhat more rigor by noting that the quantities like pressure and gravitational potential can be expected to vary quite rapidly in the z-direction in the galaxy, but should vary much more slowly in the $x - y$ plane. Thus, we can neglect derivatives of these quantities with respect to x and y. The Euler equation then becr nes

$$\frac{1}{\rho}\frac{\partial P}{\partial z}\hat{z} = \omega^2\mathbf{r} - \nabla\Omega, \tag{2.F.1}$$

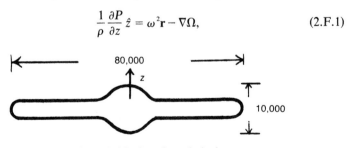

Fig. 2.7. A side view of a typical galaxy.

where **r** is the vector in the $x - y$ plane. The Poisson equation is

$$\nabla^2 \Omega \approx \frac{\partial^2 \Omega}{\partial z^2} = -4\pi G\rho. \tag{2.F.2}$$

In Problem 1.5 it was shown that equations of this type lead to a density distribution that falls off, at large z, as

$$\rho \sim e^{-z/z}.$$

Thus, most of the matter in the galaxy is located near the plane $z = 0$, and our approximation (replacing the galaxy by a disk) will be a good one.

Now consider such a disk. The Euler equation for a particle a distance r from the center (neglecting derivatives of the pressure with respect to r) is just

$$\omega^2(r) = \frac{1}{r} \frac{\partial \Omega}{\partial r}. \tag{2.F.3}$$

Thus, to find the rotational frequency (and hence the velocity) of the point at r, we need to calculate the gravitational potential at r due to the other mass elements in the disk. We do this by calculating the potential at r due to a point at r', and then adding up over all r'. (See Fig. 2.8.)

$$\Omega(r) = G \int_0^\pi \int_0^\infty \frac{M(r') r' \, dr' \, d\theta}{[r^2 + r'^2 - 2rr' \cos \theta]^{1/2}}, \tag{2.F.4}$$

where $M(r')$ is the mass per unit area at the point r', and the quantity that appears in the denominator of the integrand is just the distance $|\mathbf{r} - \mathbf{r'}|$.

Recalling that the velocity of a point in the disk depends on $\partial \Omega/\partial r$, we see that the form of rotation curve that a given galaxy will have depends very strongly on $M(r)$, the distribution of mass in the galaxy. In Problem 2.2, for example, we show that a disk with a uniform mass distribution leads, at least at small r, to solid body rotation. Let us examine some other

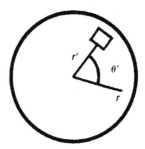

Fig. 2.8.   Coordinates for a volume element in a rotating galaxy.

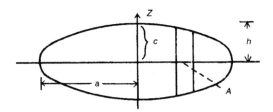

Fig. 2.9. The mass distribution derived from a Maclaurin ellipsoid.

simple examples to see what conclusions we can draw about the relation between the mass distribution in a galaxy and its rotation curve.

Let us begin by asking how one would go about replacing one of our equilibrium shapes—say a Maclaurin ellipsoid—by a flat disk. If we take an ellipsoid and imagine it broken up into columns (see **Fig. 2.9**) and then imagine each column collapsed into the plane $z = 0$, but in such a way that the mass in each column would be conserved, the mass enclosed in each column would be

$$M = \rho h A = 2c\rho A \sqrt{1 - \frac{r^2}{a^2}},$$

where A is the area of the column. Thus, the mass per unit area in the disk is just

$$M(r) = M_0 \sqrt{1 - \frac{r^2}{a^2}}, \qquad (2.\text{F}.5)$$

where we have written $2c\rho = M_0$. Now we could go ahead and put this mass distribution into the potential integral in Eq. (2.F.3) and work it out. However, we already have an expression for the potential of a Maclaurin ellipsoid,

$$\Omega = \pi\rho(\alpha_0 r^2 + \gamma_0 z^2 - \chi_0).$$

Since we are dealing with a disk, we can set $z = 0$ in the above, so that the force balance equation becomes

$$\omega^2 = \frac{1}{r} \frac{\partial \Omega}{\partial r} = 2\pi\rho\alpha_0 = \text{const.},$$

which is the curve for pure solid body rotation.

In other words, if we imagine the galaxy starting out as a flattened Maclaurin ellipsoid, we would get pure solid body rotation, unlike that which is seen for a large number of galaxies, including our own. How can we understand this?

One way is to note that the mass distribution $M(r)$ in Eq. (2.F.5) is actually pretty uniform over large distances in the galaxy. On the other hand, we know that our galaxy has a core, with an appreciable percentage of its mass lying at relatively small distances from the galactic center. Such a distribution will, of course, be poorly represented by a smooth distribution of the type given in Eq. (2.F.5). Suppose we tried a distribution like

$$M(r) = \frac{\gamma}{r} \qquad (r < R),$$
$$M(r) = 0 \qquad (r > R),$$

instead. This distribution, although singular and therefore not completely reasonable, at least does have the property of making the galaxy more massive near its center. We can put this distribution into Eq. (2.F.3), and, proceeding just as before, find that

$$-\frac{1}{r}\frac{\partial \Omega}{\partial r} = \frac{2\pi G\gamma}{r^2}\left(1 + \theta\left(\frac{r^2}{R^2}\right)\right),$$

which means that

$$(r\omega)^2 = v^2 = 2\pi G\gamma. \qquad (2.F.6)$$

This, of course, is the constant velocity rotation which was discussed above.

We see, then, that different mass distributions lead to different rotation laws, and that mass distributions which place most of the mass near the center of the galaxy tend to favor constant velocity rotation, while those which are more uniform tend to favor solid body rotation.

The question of why a galaxy should assume one mass distribution instead of another is one which cannot be treated with the simple methods we have at our disposal at this point, but is an interesting problem in itself.

G. THE RINGS OF SATURN

Astronomers have puzzled over the rings of Saturn ever since they were discovered. When the science of fluid mechanics was first developed, it was natural that the question of whether they could be composed of a fluid in equilibrium should have come up. The problem can be stated as follows: Imagine a central body of mass M surrounded by an annulus of elliptical cross section rotating with frequency ω about the

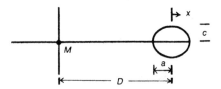

Fig. 2.10. Side view of the rings of Saturn.

body as shown in Fig. 2.10. Let us further assume that

$$\frac{a}{D}, \quad \frac{c}{D} \ll 1.$$

The potential Ω which must be inserted into the Euler equation can be written

$$\Omega = \frac{GM}{[(D+x)^2 + z^2]^{1/2}} + \Omega_R, \tag{2.G.1}$$

where the first term represents the potential at a point in the annulus due to the attraction of the central body, while the second (which we have yet to calculate) represents the potential due to the rest of the material in the annulus.

We could, of course, calculate Ω_R directly, as we did the potential for the ellipsoid, but we can get it much more easily if we note that under the conditions in Eq. (2.G.1), we can treat the annulus (at least for the purpose of calculating Ω_R) as an infinite cylinder of elliptical cross sections, whose surface is given by the equation

$$\frac{x^2}{a^2} + \frac{z^2}{c^2} = 1. \tag{2.G.2}$$

In this case,

$$\Omega = \pi\rho G(\alpha_0 x^2 + \gamma_0 z^2), \tag{2.G.3}$$

where

$$\alpha_0 = \frac{2c}{a+c}$$

and

$$\gamma_0 = \frac{2a}{a+c}.$$

The integrated Euler equation is then

$$\frac{P}{\rho} = \frac{1}{2}\,\omega^2(x+D)^2 - \pi\rho G(\alpha_0 x^2 + \gamma_0 z^2) + \frac{GM}{[(D+x)^2 + z^2]^{1/2}}$$

$$= x^2\left(\frac{1}{2}\,\omega^2 - \pi\rho G\alpha_0 + \frac{GM}{D^3}\right) - \left(\pi\rho G\gamma_0 + \frac{GM}{2D^3}\right)z^2$$

$$+ x\left(D\omega^2 - \frac{GM}{D^2}\right) + \cdots, \tag{2.G.4}$$

where we have dropped terms higher than second order in x/D and z/D.

We see immediately that unless the coefficient of the term linear in x vanishes, the surfaces of constant pressure will never coincide with the surface of the annulus. This means that

$$\omega^2 = \frac{GM}{D^3}, \tag{2.G.5}$$

i.e. that there is only one frequency at which the annulus can rotate, regardless of its shape. This is a departure from our previous results, in which an equilibrium was possible at any frequency up to ω_c. Here the frequency is completely fixed by the central body.

We note in passing that this frequency is precisely that which a satellite in orbit around the central mass would have.

In order for the surfaces of constant pressure to coincide with the surface of the ring, we must have

$$a^2\left(\frac{\omega^2}{2} - \pi G\rho\alpha_0 + \frac{GM}{D^3}\right) = -c^2\left(\gamma_0 G\pi\rho + \frac{GM}{2D^3}\right),$$

which gives the equilibrium condition as

$$\frac{\omega^2}{\omega_c^2} = \frac{2ac(a-c)}{(3a^2 + c^2)(a+c)}. \tag{2.G.6}$$

Thus, provided that the ratio a/c can be adjusted to satisfy this condition (where ω is no longer free, but determined by Eq. (2.G.5)), the rotating ring will be in equilibrium.

Have we, then, found the solution to the problem of the composition of the rings of Saturn? Unfortunately, the answer to this question is no. Up to this point in the text, we have considered only the question of whether or not a fluid mass could be in equilibrium. But there are both unstable and stable equilibria, and it turns out that the one treated in this section is of the former variety. In Problem 3.2, it is shown that a small perturbation of the center of the ring will lower the energy of the ring system, so that a fluid ring of the type we have discussed would not survive long in nature.

The concept of stability is, however, a very important one in fluid mechanics, and we will now turn to a full discussion of it.

SUMMARY

We have seen that by going to a frame rotating with a fluid mass, the dynamical problem of calculating the motion of such fluids can be replaced by the static problem of balancing pressure, centrifugal force, and gravitation. The method of calculating equilibrium shapes for such bodies is quite simple in principle (although sometimes complicated mathematically). We simply calculate the gravitational potential for the body, insert this into the Euler equation, and demand that a surface of constant pressure coincide with the surface of the body. In this way, various physical systems were examined, including ellipsoids (such as the earth), disks (such as the galaxy), and rings (such as those around Saturn), and it was found to be possible to find equilibrium configurations for each shape.

PROBLEMS

2.1. Show that, unlike the earth, the approximation of treating the sun as a rotating ideal fluid does not give good agreement between theory and observation for ω/ω_c. The fact that the outer surface of the sun rotates slowly has caused many problems in astrophysics.

2.2. Consider a galaxy which has a mass distribution given by

$$M(r) = M_0 \qquad (r < R_0),$$
$$M(r) = 0 \qquad (r > R_0).$$

Show that this leads to an expression for angular frequency given by

$$\omega^2(r) = \frac{4GM_0}{R_0} \int_0^{\pi/2} \frac{\sin^2 \psi \, d\psi}{[1 - (r^2/R_0^2) \sin^2 \psi]^{1/2}}.$$

(*Hint*: You may find the following change of coordinates useful

$$r' \sin \theta' = s \sin \psi,$$
$$r - r' \cos \theta' = s \cos \psi.)$$

Hence, show that in the limit $r/R_0 \to 0$, this distribution gives a solid body rotation just like the Maclaurin ellipsoid. (*Hint*: You might want to consult L. Mestel, *R.A.S. Monthly Notices* **126**, 553 (1963).)

2.3. Show that it is impossible for a prolate spheroid to be in equilibrium. This corresponds to our intuition, which tells us that centrifugal force will tend to pull a rotating body out at the equator, thus leading to oblate shapes.

2.4. Prove that a rotating body in equilibrium must be symmetrical about a plane through its center and perpendicular to the axis of rotation. (*Hint*: Show that if this were not true, the pressure at the points on the surface at the tips of a column through the fluid perpendicular to the plane could not be equal.)

2.5. A very serious problem in astronomy is determining how much matter there is in the universe, since not all matter is luminous and therefore visible. For example, we know that there is a lot of dust in the galaxy which can be detected only by looking at light which has come through it. Suppose in a distant galaxy we observed a density of luminous matter

$$M_L(r) = M_0 \sqrt{1 - \frac{r^2}{R^2}},$$

where R is the radius of the galaxy. Suppose we also observed a rotation curve given by

$$V = cr.$$

Find an expression for η, the ratio of luminous to nonluminous matter from these experimentally determined numbers.

2.6. Calculate the angular momentum of the sun and of the entire solar system. Which bodies carry most of the angular momentum?

2.7. The theories of the sun's formation which are now accepted suggest that the sun condensed out of a gas which was initially rotating. As an example of this process, consider a sphere of gas of mass M and angular momentum L. Suppose that this sphere collapses by some process which we do not follow to a Maclaurin ellipsoid whose major axis is of length a.

(a) If we conserve M and L, write an expression for the density of the gas in the final state as a function of the parameter ξ.

(b) Hence write one (complicated) equation for ξ itself.

(c) For a body like the sun, which is nearly spherical, solve for ξ and hence ω, the frequency of rotation.

(d) If the original cloud was the size of the solar system, how much did the sun speed up when contracting?

2.8. Using the methods of Appendix B, find the electrostatic potential at the points inside an ellipsoid which has a charge σ per unit volume.

2.9. Consider an ellipsoid which has a charge density σ per unit volume and a matter density ρ per unit volume.

(a) Derive the expression corresponding to Eq. (2.A.4) for such a system.

(b) Hence find the surfaces of constant pressure, and write down the condition which tells whether the ellipsoid can be in equilibrium.

(c) Define a new critical frequency for the charged ellipsoid. Can it ever be zero? Give a physical interpretation of this result.

2.10. It has sometimes been suggested that the galaxies are moving away from each other because of a small electrostatic charge on each galaxy. How would Eq. (2.F.3) be changed if this were so? Under what conditions would the galactic rotation curve for a charged and uncharged galaxy be the same?

2.11. Calculate the critical frequency of the earth, the sun, and as many of the planets as you can. Are any near this limit?

2.12. One theory for the formation of the asteroid belt (which is not accepted today) is that the asteroids are the result of the disruption of a planet. Let us call this planet Krypton for definiteness, and argue as follows: Since the planet was near the earth and Mars, it was presumably formed in the same way, and hence should rotate with about the same speed. If this were so, what would its density have to be to have it disrupt because of the mechanisms discussed in this chapter? Are there any materials of this density known? Are they prevalent in the asteroid belt?

REFERENCES

A discussion of stability of gravitating fluids is given in H. Lamb, *Hydrodynamics* (cited in Chapter 1). For more detailed presentations of the principles of stellar structure, see

S. Chandrasekar, *An Introduction to the Study of Stellar Structure*, Dover Publications, New York, 1957.

John P. Cox and R. T. Giugli, *Principles of Stellar Structure*, Gordon and Breach, New York, 1968.

3

The Idea of Stability

Bright star, were I as steadfast as thou art!

JOHN KEATS
Sonnet written on a blank page
in Shakespeare's poems

A. INTRODUCTION

Up to this point we have only been concerned with questions related to the possibility of balancing forces in fluid masses. We have, in other words, looked only for situations in which it was possible to establish equilibrium. We have not asked whether the equilibrium configurations which we have found were stable. For a system to be in stable equilibrium, we must not only have a situation in which forces are in balance, but where small deviations of the system from the equilibrium must generate forces which tend to drive the system back toward its equilibrium configuration, rather than farther away from it. The classic example of such a system is a mass on the end of an unstretched spring. Any movement of the mass away from this equilibrium position results in the spring exerting a force pulling (or pushing) the mass back toward its original position.

A ball sitting on top of a hill would be an example of a unstable equilibrium, since small changes of position would result in the ball being driven farther and farther from equilibrium. A third type of equilibrium—neutral equilibrium—can be defined between these two. This is a situation in which movement away from the equilibrium position results in no forces being exerted at all. A ball on a flat table top would be an example of such a system. Let us now try to formulate these ideas more quantitatively.

39

Let us begin with a system in which the kinetic energy can be neglected, and where the potential energy can be written as $V(q_1 \ldots q_i \ldots)$ where the q_i are some coordinates. Then the condition for equilibrium is that the forces on the system cancel—i.e. that at equilibrium

$$\frac{\partial V}{\partial q_i}\bigg]_0 = 0. \tag{3.A.1}$$

To investigate stability, we must ask how the system behaves if we move slightly away from equilibrium—i.e. if we let

$$q_i \rightarrow q_{i0} + \eta_i. \tag{3.A.2}$$

To make such an investigation, we expand the potential near the equilibrium point in a Taylor series

$$
\begin{aligned}
V(q_1 \ldots q_i \ldots) &= V(q_{10} \ldots q_{i0} \ldots) \\
&+ \sum_i \frac{\partial V}{\partial q_i}\bigg]_0 \eta_i \\
&+ \frac{1}{2} \sum_{i,j} \frac{\partial^2 V}{\partial q_i \partial q_j}\bigg]_0 \eta_i \eta_j + \cdots.
\end{aligned}
\tag{3.A.3}
$$

Now from the equilibrium condition, the term linear in the displacement η_i is zero, so we see that the change in V as we move away from equilibrium is governed by the sign of the second derivatives of the potential. If the term bilinear in η is positive, then moving away from equilibrium tends to increase the energy of the system, so that the equilibrium is stable. If this term is negative, however, then small deviations tend to decrease the energy of the system, and the equilibrium will be unstable.

To fix these ideas more firmly, let us consider the case of a potential which depends on only one coordinate q and on one other parameter λ. In the example of the particle on the spring, these would be the position of the particle and the spring constant. In this simple example, the potential as a function of the two parameters q and λ would be a surface in three dimensions. An example of such a surface is pictured in Fig. 3.1.

To begin our discussion of stability, let us consider only those perturbations in which q is varied while λ is held fixed. We will consider other types of perturbations later.

At the point P, a plane at fixed λ gives a curve of V versus q which looks like Fig. 3.2. Thus, either by inspection or from Eq. 3.A.3, we see that the system is stable at the point P against perturbations in which λ is held fixed.

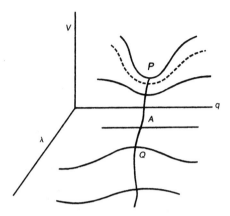

Fig. 3.1. Potential surface as a function of two parameters.

In terms of our spring example, at this point, the spring will tend to pull the system back into equilibrium.

At the point Q, the situation is somewhat different. Here the cut through the potential surface at constant λ yields a graph like Fig. 3.3, so that the system is unstable against perturbations with constant λ at this point.

The transition between these two cases occurs at A, where the potential surface looks like Fig. 3.4. This represents neutral equilibrium, where the second derivatives of the potential vanish, so that displacements of the system do not change its energy at all.

Thus, the potential surface we have drawn as an example illustrates all of the types of stability discussed earlier. It also illustrates another very important point about stability. To see this point, let us go back to our consideration of the point P. Previously, we had considered only those perturbations in which we changed q slightly, but held λ fixed. Let us now consider the other alternative—let us consider a perturbation in which q is held fixed and λ is varied (think, for example, of holding the position of the

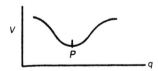

Fig. 3.2. Potential surface at P for fixed λ.

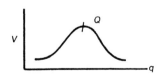

Fig. 3.3. Potential surface at Q for fixed λ.

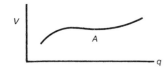   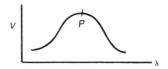

Fig. 3.4. Potential surface at A for fixed λ.   **Fig. 3.5.** Potential surface at P for fixed q.

particle at the end of a spring fixed, but heating the spring so that the spring constant changes). At P, this corresponds to looking at a plane perpendicular to the q-axis, in which case we have Fig. 3.5.

In other words, the system at P was stable against the first type of perturbation but unstable against the second! This is a very important point when discussing stability—one must always specify against which types of perturbation the system is stable. There are many systems (we shall consider one in the next section) which are stable against one type of perturbation, but unstable against another. In our two-dimensional example, then, we would say that a point was a point of stability if and only if the second derivatives of V were positive in every direction around the point, or, equivalently, the potential exhibited a minimum in every possible plane drawn through the point.

If this were not the case, small thermal fluctuations would eventually move the system slightly in the direction in which the potential would be lower, and, once started, nothing could bring it back (this is similar to a ball rolling down a hill—it takes only a small push to start it going).

Our example has concerned itself only with a potential which depends on two variables. In general, potentials will depend on many more variables than this. For example, a particle moving in three dimensions attached to three springs would depend on six variables—the x, y, z coordinates of the particle and the three spring constants. The potential would then be a surface in a seven-dimensional space. The idea of finding minima and maxima, and the other properties of stability discussed for the simple example above, however, is still applicable, and provides a useful way to visualize the problem.

Let us now turn to a discussion of the evolution of systems in time. We go back to our simple two-dimensional example, and suppose that now we deal with a piece of the potential surface which looks like Fig. 3.6.

The line XYZ now represents a line of extrema of the surface. The point Z represents a state of the system which is unstable against any perturbation, while the point X represents a truly stable state. Now if we start the system off at some point L, which is not necessarily a point of equilibrium or stability, the system will evolve in time, just as a ball placed on the side of a

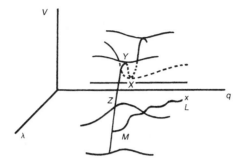

Fig. 3.6. The evolution of a system along a potential surface.

hill will start rolling. Let the line LM represent the states through which the system passes (it might be helpful to visualize this in terms of a particle on a spring—when the spring constant is changed by heating, for example, the position of the particle will change. This leads to new values of λ and q, and hence to a new state of the system, represented by a new point on the surface).

If the point M happens to fall on the curve XYZ between X and Y, then the system has a chance of achieving stability, while if it falls between Y and Z, it does not. (Again, thinking of the motion of the system as a ball rolling around on the potential surface will help to visualize this point.)

It is interesting to ask what happens if the point M falls exactly on the point of neutral equilibrium Y. In this case, the system can "choose" stability or instability. The situation is similar to balancing a ball on a point and asking which way it will fall. The answer depends on a large number of factors—the precise way in which the ball was placed, slight movements of the air or vibrations of the floor, etc. Such effects, while calculable in principle, are usually regarded as random factors beyond the range of analysis. But it is clear that at the point Y, a slight displacement of the system toward Z will result in instability of the system, while a slight deviation toward X will result in stability.

B. STABILITY OF THE MACLAURIN ELLIPSOID

As an example of the discussion of stability in the previous section, we examine the Maclaurin ellipsoid's stability against a certain type of perturbation. Before doing so, however, we have to remember that when we are dealing with a rotating gravitating fluid, the energy is made up both of kinetic and potential contributions, rather than just potential energy, as it was in the simplified model we considered in the previous section. Thus, the condition for an equilibrium becomes

$$\frac{\partial}{\partial q_i}(T + V) = 0, \tag{3.B.1}$$

while the condition for stability is

$$\frac{\partial^2}{\partial q_i \partial q_i}(T + V) > 0. \tag{3.B.2}$$

These new conditions correspond to the fact that every system will tend to move toward a state of lowest total energy.

We shall consider a very restricted class of perturbations: those perturbations which

    (i) conserve angular momentum,
    (ii) preserve that geometry of the Maclaurin ellipsoid (i.e. those perturbations which keep two axes equal),
    (iii) keep the density constant.

The first restriction is very reasonable if we think of things like stellar bodies, since any perturbation in such a system has to come from within the system itself, and hence preserve angular momentum. The second restriction will be imposed during the course of the discussion for mathematical simplicity.

The kinetic energy of the system in terms of the angular momentum L is just

$$T = \frac{L^2}{2I}, \tag{3.B.3}$$

where I is the moment of inertia about the axis of rotation and is given by

$$I = \frac{1}{5}M(a^2 + b^2).$$

The potential can be calculated in a straightforward manner (see Problem 3.1) to be

$$V = \frac{1}{2}\int \rho \Omega \, dV = -\frac{8\pi}{15}\rho abc \int_0^\infty \frac{d\lambda}{\Delta}. \tag{3.B.4}$$

where the symbols are defined in Chapter 2.

Let us begin by noting that the question of stability of an ellipsoid now comes down to finding minima in the function $E = T + V$. In general, this is a function of a, b, and c. However, requirement (iii) means that if a and b are changed, the requirement of constant volume then determines the value of c. Thus, E will be considered to be a function of a and b only. Later, we shall impose restriction (ii) and consider the case $a = b$ only. For the moment, however, let us keep the more general case under consideration.

We could, of course, calculate the value of E for every value of a and b and look for minima. We can get an answer in the case of the Maclaurin ellipsoid without such a complicated procedure, however. Write

$$E(a, b) = T + V.$$

Now as $a \to \infty$, we would have a situation in which the material in the ellipse was spread out over all space, so we would expect $V \to 0$. Clearly, in this limit $T \to 0$ [see Eq. (3.B.3)] as well, so that $E \to 0$. A similar argument holds for the limit $b \to \infty$.

From the expression for V in Eq. (3.B.4), we see that if either $a \to 0$ or $b \to 0$, $V \to 0$. If $a \to 0$, then $E(a, b) \propto 1/b^2$ (similarly, if $b \to 0$, $E(a, b) \propto 1/a^2$). This means that the function $E(a, b)$ must look like Fig. 3.7 in the regions dealt with above.

If we restrict our attention to the Maclaurin ellipsoid, we want only the plane containing the line $a = b$. Note that by restricting our attention to this plane, we are only considering stability against perturbations which leave the lengths of the two major axes equal, and we will be unable to say anything about perturbations which change these lengths differently. However, in this plane, the function $E(a, b)$ can be sketched out. We know that it must (1) become infinite as $a = b \to 0$, (2) go to zero as $a = b \to \infty$, and (3) from Section 2.C, we know that there is one and only one point of equilibrium— i.e. only one point at which $\partial E/\partial a = 0$. This means that $E(a, b)$ in this case must look like Fig. 3.8, which means that the Maclaurin ellipsoid is stable

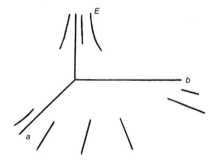

Fig. 3.7. The energy for a Maclaurin ellipsoid as a function of a and b.

Fig. 3.8. The energy surface along the line $a = b$.

against perturbations in which $a = b$. From the argument so far, we can draw no conclusions about the stability against other types of perturbations.

In fact, the Maclaurin ellipsoid is stable against all perturbations involving bulk changes of the relative size of the axes, as are the Jacobi ellipsoids. This means that the minimum in the $a = b$ plane shown above is actually a minimum in the surface $E(a, b)$, and not a saddle point. Other minima in the surface would correspond, of course, to the Jacobi ellipsoids.

For completeness, it should be noted that these ellipsoids, while stable against perturbations which leave the density of the fluid unchanged, are unstable against fluctuations in this density. This illustrates the point which was made earlier—that it is possible for a system to be stable against one type of perturbation but not against another.

SUMMARY

The question of the stability of a fluid system was discussed. The general requirement that a system be in stable equilibrium is that every possible perturbation of the system lead to a state of higher total energy. It is always possible, of course, that a system could be stable against one type of perturbation, but unstable against another. The stability of the Maclaurin ellipsoid was investigated, and it was shown that the equilibrium configurations derived in the previous chapter were indeed stable against perturbations which keep the density of the fluid constant.

PROBLEMS

3.1. Given the expression for the potential inside of an ellipsoid from Appendix B, find the total gravitational potential energy of such a body, and hence verify Eq. (3.B.4).

3.2. A full discussion of the stability of the rings of Saturn would be a long undertaking. However, there is a relatively simple calculation that can be done to show the instability of the rings if we assume that the rings are solid (clearly, if the rings cannot be stable if solid, they are unlikely to be stable if they are fluid). Consider a solid ring of circular cross section a, mass m, and radius D centered on an attracting body of mass M. Show that if the center of the ring is displaced slightly from the center of the attracting body, the energy of the system is lowered, so that the system is unstable.

3.3. Consider a Maclaurin ellipsoid of mass density ρ and charge density σ.

(a) Calculate the total potential energy in such a system, including both electrical and gravitational contributions.

(b) Under what conditions will such an ellipsoid be stable? (*Hint*: You may wish to refer to Problem 2.8.)

3.4. Consider a situation as shown in Fig. 3.9, in which a particle at the point (L, L) is attached to two springs of equal spring constants k and unstretched length L.

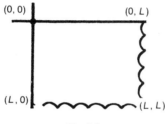

Fig. 3.9.

(a) Calculate the potential energy of the system if the particle is moved to an arbitrary point (X, Y).

(b) Are there any other points of equilibrium in the plane?

(c) Are these points stable or unstable equilibria?

3.5. Repeat the analysis of Problem 3.4 for the case when the particle carries a charge q, and a charge Q (of the same sign) is located at the origin.

3.6. An interesting kind of instability is occasionally encountered in dealing with binary star systems. Consider two stars, of mass m and m', located a distance R apart and rotating about the common center of mass with frequency ω.

(a) Give an argument leading to the conclusion that

$$\omega^2 = \frac{G(m + m')}{R^3}.$$

(b) Show that the potential at any point in space is given by

$$\Omega = \frac{Gm}{r} + \frac{Gm'}{r'} + \frac{\omega^2}{2}\left\{\left(X - \frac{m'R}{m + m'}\right)^2 + Y^2\right\},$$

where the arbitrary point is (X, Y, Z), and r and r' are the distances from the masses to the point.

(c) Show that if we define

$$X = r\lambda,$$
$$Y = r\mu,$$
$$Z = r\nu,$$

and

$$\rho = \frac{r}{R},$$

$$q = \frac{m'}{m},$$

the potential becomes

$$\Omega = \frac{1}{\rho} + q\left[\frac{1}{(1 - 2\lambda\rho + \rho^2)^{1/2}} - \lambda\rho\right] + \frac{q+1}{2}\rho^2(1 - \nu^2).$$

(d) Make a sketch of the potential in part (c) for various values of q. Produce an argument that for some value of q, it should be possible for a particle to go from the gravitational field of one star to that of the other without expending energy. When this happens, we speak of having reached *Roche's limit*, in which mass will be exchanged between the stars.

REFERENCES

For a general discussion of the stability of physical systems, see

Robert A. Becker, *Introduction of Theoretical Mechanics*, McGraw-Hill, New York, 1954 (Chapter 5).

S. Chandrasekar, *Hydrodynamics and Hydromagnetic Stability*, Clarendon Press, Oxford, 1961.

4
Fluids in Motion

No man steps into the same river twice.

A. THE VELOCITY FIELD

Up to this point, we have been considering only the case of hydrostatics, which deals with stationary fluids. Even the case of rotating stars was treated by going to a rotating frame of reference, in which the fluid which comprised the star would not be in motion. We now turn our attention to the more general case of moving fluids, the study of hydrodynamics.

The first thing which we shall have to decide is how to characterize the motion of the fluid. If we think of the fluid as being composed of infinitesimal volume elements, then a volume element located at coordinates (x, y, z) will have some velocity $\mathbf{v}(x, y, z, t)$ (see Fig. 4.1). This means that to each point in space we can assign a vector which can be, in general, a function of both position and time. This collection of velocities is referred to as a *velocity field*.

It is possible to write down the velocity at the point (x, y, z) in terms of the velocity vector and its derivatives at the origin by using a Taylor expansion

$$v_i(x, y, z) = v_i(0, 0, 0) + \sum_j \frac{\partial v_i}{\partial x_j} x_j + \cdots. \tag{4.A.1}$$

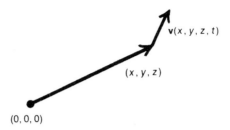

Fig. 4.1. The velocity field.

If we confine our attention to a small neighborhood near the origin, so that x, y, and z are small, we can ignore higher-order terms in this expansion and express the velocity field near the origin in terms of the derivatives of the velocity. Eq. (4.A.1) can be written in the form

$$v_i(x, y, z) = v_i(0, 0, 0) + \frac{\partial v_i}{\partial x_i} x_i + \sum_{j \neq i} \frac{\partial v_i}{\partial x_j} x_j$$

(in this equation, the summation convention is not used). By adding and subtracting the same thing to the term inside of the summation, this can be cast in the form

$$v_i(x, y, z) = v_i(0, 0, 0) + \frac{\partial v_i}{\partial x_i} x_i$$
$$+ \sum_{i \neq j} \frac{1}{2} \left[\frac{\partial v_i}{\partial x_j} + \frac{\partial v_j}{\partial x_i} + \frac{\partial v_i}{\partial x_j} - \frac{\partial v_j}{\partial x_i} \right] x_j. \qquad (4.A.2)$$

Thus, the change in velocity as we move from one point in the fluid to another can be written as the sum of three parts,

$$v_i(x, y, z) - v_i(0, 0, 0) = \Delta v_i$$
$$= D_i + S_i + C_i, \qquad (4.A.3)$$

where

$$D_i = \frac{\partial v_i}{\partial x_i} x_i \qquad (4.A.4)$$

is related to the divergence of the velocity field (i.e. $\partial v_i / \partial x_i$ is one piece of the divergence $\nabla \cdot \mathbf{v}$),

$$C_i = \frac{1}{2} \sum_{j \neq i} \left(\frac{\partial v_i}{\partial x_j} - \frac{\partial v_j}{\partial x_i} \right) x_j \qquad (4.A.5)$$

is related to the curl of the field, and the remaining term,

$$S_i = \frac{1}{2} \sum_{j \neq i} \left(\frac{\partial v_i}{\partial x_j} + \frac{\partial v_j}{\partial x_i} \right) x_j, \tag{4.A.6}$$

will just be called the "symmetric part."

The purpose of writing Δv_i in this rather cumbersome way is to try to understand what different sorts of velocity field correspond to in terms of physical movement of the fluid. For example, we shall see that there is an intimate relationship between the expression $\nabla \times \mathbf{v}$ and rotational motion in the fluid, and between the expression $\nabla \cdot \mathbf{v}$ and changes of density. Thus, it will be possible to go from the rather formal definition of a velocity field which we have given above, in which each point in space is associated with a vector, to a physical picture of what sort of fluid motion is associated with velocity fields with different kinds of properties.

The technique which we shall use to accomplish this will be to consider four points in the fluid at time $t = 0$ (see Fig. 4.2). We shall then compute the velocity at each corner of the square in terms of D, S, and C, which shall be calculated from the given velocity field itself. We shall then ask what the four points look like an infinitesimal time τ later. Each point will have moved a certain infinitesimal distance. For example, the point $(0, L)$ will have moved a distance

$$\Delta x = v_x(0, L)\tau$$

in the x-direction, and a distance

$$\Delta y = v_y(0, L)\tau$$

in the y-direction. Similar results will be obtained for each of the other points, so that (restricting our attention to two-dimensional flow), at time τ, we shall have the situation in Fig. 4.3. Thus, having calculated D, S, and

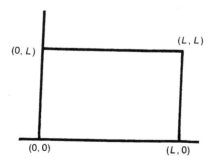

Fig. 4.2.   The initial square in a moving fluid.

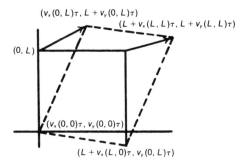

Fig. 4.3. The final configuration of the square.

C from the velocity field, we can immediately visualize the type of motion which is being executed by the fluid.

Of course, we could do this directly, without calculating D, S, and C, by taking the velocities at the points of the square directly from the velocity field. We shall see in later sections, however, that the divergence and the curl of the velocity field play a special role in describing fluid flow, and hence it is important to describe fluid motion in the way we have above.

We shall proceed by looking at three examples, in which velocity fields are chosen so that only one of the three terms in Δv_i is nonzero.

Example I

Consider a velocity field in two dimensions given by

$$v_x = Cx,$$
$$v_y = 0. \tag{4.A.7}$$

This will result in a velocity configuration like that shown in Fig. 4.4. For this field, we have

$$D_x = \frac{\partial v_x}{\partial x} x = Cx$$

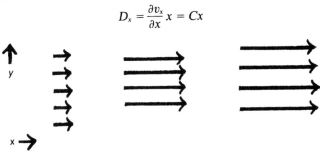

Fig. 4.4. The velocity field for Example I.

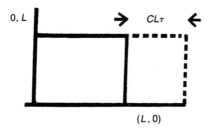

Fig. 4.5. The final configuration for Example I.

and

$$C_i = S_i = D_y = D_z = 0.$$

Thus, the x-components of the velocity at each of the four points are given by

$$v_x(0, 0) = 0,$$
$$v_x(0, L) = 0,$$
$$v_x(L, L) = CL,$$
$$v_x(L, 0) = CL.$$

(4.A.8)

The square at time τ will then appear as in Fig. 4.5.

Thus, a velocity field which possesses a nonzero divergence will give rise to motion which can be characterized as a stretching along one of the major axes. This is quite a reasonable result, since we know that for an incompressible fluid, the equation of continuity becomes

$$\nabla \cdot \mathbf{v} = 0,$$

so that the existence of a divergence implies that there must be a changing density in order for continuity to be satisfied. Pictorially, we see that such a change of density must occur, too, since the area bounded by the lines in the above figure changes, but no fluid crosses the boundaries, so that the density must decrease.

Example II

Consider a velocity field given by

$$v_x = Cy,$$
$$v_y = Cx.$$

(4.A.9)

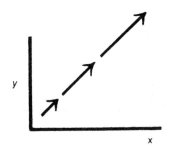

Fig. 4.6.   The velocity field for Example II.

This will result in a velocity configuration like that shown in Fig. 4.6, which has

$$S_x = Cy,$$
$$S_y = Cx,$$
$$C_i = D_i = S_z = 0.$$

For such a field, the x- and y-components of velocity at the points of the square are

$$v_x(0, 0) = 0 = v_y(0, 0),$$
$$v_x(L, 0) = 0 = v_y(0, L),$$
$$v_x(0, L) = CL = v_y(L, 0),$$ (4.A.10)
$$v_x(L, L) = CL = v_y(L, L).$$

The square at time τ will then appear as in Fig. 4.7.

We see, then, that a velocity field characterized by a nonzero symmetric part also corresponds to a uniform stretching of the fluid, but this time along some axis other than a coordinate axis.

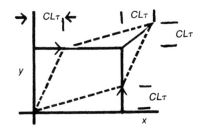

Fig. 4.7.   The final configuration for Example II.

Example III

As a final example, consider a velocity field given by

$$v_x = Cy,$$
$$v_y = -Cx. \tag{4.A.11}$$

This will result in a velocity configuration like that shown in Fig. 4.8, and has

$$D_i = S_i = 0,$$
$$C_x = C \cdot y,$$
$$C_y = -C \cdot x.$$

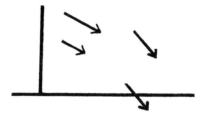

Fig. 4.8.   The velocity field for Example III.

The velocities of the corners of the square are now

$$v_x(0, 0) = 0 = v_y(0, 0),$$
$$v_x(L, 0) = 0 = v_y(0, L),$$
$$v_x(0, L) = CL = v_x(L, L), \tag{4.A.12}$$
$$v_y(L, 0) = -CL = v_y(L, L).$$

The square at time τ will then appear as in Fig. 4.9.

Simple geometry shows that this velocity field corresponds to a rotation of the square around the origin, with no change in area. Thus, the existence of

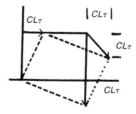

Fig. 4.9.   The final configuration for Example III.

the curl of a velocity field corresponds to rotational motion, just as the existence of a divergence of a symmetric part corresponds to stretching motion.

With this understanding, we can now look at some general features of fluid flow.

B. THE VELOCITY POTENTIAL

We have seen that if we know the velocity of the fluid elements in a volume as a function of the position, we can make some very general statements about the type of fluid motion which occurs. In particular, if the field is such that

$$\nabla \cdot \mathbf{v} = 0, \tag{4.B.1}$$

the fluid density cannot change, and only motions which conserve density are allowed. On the other hand, if the field is such that

$$\nabla \times \mathbf{v} = 0, \tag{4.B.2}$$

then no rotational motion is allowed in the fluid. For obvious reasons, such a flow is called *irrotational flow.*

It is clear that if we have irrotational flow, the velocity can be written

$$\mathbf{v} = \nabla \phi, \tag{4.B.3}$$

where ϕ is a scalar function called the *velocity potential.* (This is very similar to the definition of a magnetic scalar potential in electromagnetic theory in the static case where $\nabla \times \mathbf{B} = 0$; so $\mathbf{B} = \nabla \phi_m$.) Thus, irrotational flow is sometimes referred to as *potential flow.* Almost all of the examples which we shall consider will involve potential flow, which is fortunate, since the introduction of a velocity potential allows us to work with scalar rather than vector quantities.

An interesting result can be written down in the special case of potential flow of an incompressible fluid. From Eqs. (4.B.1) and (4.B.3), we have

$$\nabla^2 \phi = 0, \tag{4.B.4}$$

which is just Laplace's equation. We shall have repeated recourse to this result in future examples.

Let us write down the Euler equation in terms of the velocity potential. If we start with the Euler equation in the form [see Eq. (1.B.5)]

$$\frac{\partial \mathbf{v}}{\partial t} + \frac{1}{2} \nabla v^2 - \mathbf{v} \times (\nabla \times \mathbf{v}) = -\frac{1}{\rho} \nabla P - \nabla \Omega,$$

then the substitution of $\nabla\phi$ for **v** yields (recalling that the curl of the gradient vanishes)

$$\nabla\left(\frac{\partial\phi}{\partial t} + \frac{1}{2}v^2 + \frac{P}{\rho} + \Omega\right) = 0, \tag{4.B.5}$$

so that, in general,

$$\frac{\partial\phi}{\partial t} + \frac{1}{2}v^2 + \frac{P}{\rho} + \Omega = f(t), \tag{4.B.6}$$

where $f(t)$ is an arbitrary function of time, and plays the role of an integration "constant." To deal with the function $f(t)$, we need to notice an important property of the velocity potential. If we have a potential ϕ which gives rise to a velocity field **v**, then any potential of the form

$$\phi' = \phi + \int^t f(t')\,dt' \tag{4.B.7}$$

will give rise to exactly the same velocity field. Since it is only **v** which can be measured, we can always add or subtract any function of time to any velocity potential without changing any of the physics of the problem. This is completely analogous to the fact that we can always add a constant term to a gravitational potential without changing any forces, and corresponds to the freedom to pick the zero of a potential wherever we like, since only potential differences can be measured. Therefore, without loss of generality, we can write

$$\frac{\partial\phi}{\partial t} + \frac{1}{2}v^2 + \frac{P}{\rho} + \Omega = \text{const.} \tag{4.B.8}$$

If, in addition to being irrotational, the flow has achieved steady state (i.e. a situation where the velocity at any given point does not depend explicitly on the time, although it may vary from point to point), then $\partial\phi/\partial t = 0$, and this reduces to

$$\frac{1}{2}\rho v^2 + P + \rho\Omega = \text{const.}, \tag{4.B.9}$$

which is a special case of the *Bernoulli equation*. In Problem 4.4, the problem of showing that the quantity

$$\frac{1}{2}\rho v^2 + P + \rho\Omega$$

is the same everywhere along a streamline in the fluid is given. This is the most general form of the Bernoulli equation, and states that while this quantity must be conserved along a streamline, it can, in general, have different values for different streamlines. For the special case of irrotational motion, however, we have shown that this quantity must not only be conserved along a given streamline, but must be the same for every streamline in the fluid.

C. STABILITY OF FLOW

In Chapter 3, we saw that a very important property of fluid systems in equilibrium was stability. We saw that for static or semi-static systems, this could be understood in terms of the properties of the energy surface. If the system was one in which the energy increased as we moved away from equilibrium, then it was stable, while if the energy decreased, it was unstable.

This same sort of reasoning can be applied to fluid flow patterns as well, although it is usually more convenient to make the calculations which allow us to decide whether a system is stable or unstable in a different way. To understand this new line of attack and connect it to the discussion of Chapter 3, let us consider the case of a ball rolling on a surface (see Fig. 4.10).

In Chapter 3, we would describe the situation on the left as unstable because as we move away from equilibrium, the energy of the system is lowered. The situation on the right, however, would be stable, since movement away from the equilibrium configuration raises the total energy.

The new reasoning which we shall apply to the fluid flow problem is as follows: In the left-hand diagram, a small displacement of the system from equilibrium will result in the ball moving far away from the top of the hill (since a small displacement will cause it to roll down). Thus, the equations of motion of the system must be such that if I allow small, time-dependent departures from equilibrium (the origin of these small perturbations is discussed in Chapter 3), then $x(t)$, the position of the ball, will eventually become quite large. For the stable configurations, however, the equations of motion are such that $x(t)$ stays small (typically, the system will perform small-scale oscillations around the equilibrium point).

The advantage of this technique is that it allows us to determine the question of stability directly from the equations of motion, without

Fig. 4.10.   The idea of stability.

calculating energy at all. For example, if we assumed that $x(t)$ was of the form

$$x(t) \sim e^{i\omega t}, \tag{4.C.1}$$

then for the stable case, when we solved the equations of motion for ω, they would require that ω be real. For the unstable case, however, they would require that ω be complex, and of the form

$$\omega = \omega_R - i|\omega_I|, \tag{4.C.2}$$

so that the time dependence of $x(t)$ would be

$$x(t) \sim e^{i\omega_R t} e^{|\omega_I| t}, \tag{4.C.3}$$

and the system would indeed "run away" when a small perturbation was applied.

To see how this idea works in the case of a fluid, let us consider the "tangential instability" in fluid flow. Let there be two fluids, of density ρ_1 and ρ_2, with the upper fluid moving with velocity $\mathbf{v}_0$. At the plane $z = 0$, the two fluids meet, and the equilibrium configuration is obviously the case where the interface between the two fluids is simply the $z = 0$ plane. The question which we can ask concerns the stability of the equilibrium. If we distort the interface slightly, will the distortions tend to smooth out or stay small, or will they grow and disrupt the flow?

To answer this question, let us assume that the surface is slightly distorted, and let ξ (see Fig. 4.11) measure the deviation of the surface from its equilibrium position. When the surface is distorted, all of the other variables in the system will change by a small amount as well, so that we will have

$$\begin{aligned}
P_1 &= P_{10} + P_1', \\
P_2 &= P_{20} + P_2', \\
v_1 &= v_0 + v_1', \\
v_2 &= v_2',
\end{aligned} \tag{4.C.4}$$

where P_1 and P_2 are the pressures in the regions of fluids 1 and 2. The subscript "0" refers to the equilibrium pressure and the P' terms are the

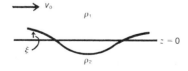

Fig. 4.11. The deformation at the interface between two fluids.

small changes in the equilibrium value caused by the small distortion of the surface. Similarly, v_1 and v_2 refer to the fluid velocities, although in this case the equilibrium velocity in region 2 is zero.

In general, all of the small quantities in Eq. (4.C.4) are complicated functions of the position and time. However, we know that each can be expanded in a Fourier series, each component of which has a behavior $e^{i(kx-\omega t)}$.

Therefore, without loss of generality, we can consider only the case

$$\xi(x, y, z, t) = \xi(z)e^{i(kx-\omega t)},$$
$$P'(x, y, z, t) = P'(z)e^{i(kx-\omega t)}, \tag{4.C.5}$$
$$\mathbf{v}'(x, y, z, t) = \mathbf{v}'(z)e^{i(kx-\omega t)},$$

since any more complicated functions of x and t can be expressed as a series of terms of this type. As we shall see, our final working equations will be linear, so if we find a solution for the general term in such a sum, the final result will simply be a sum of such solutions (see Appendices E and F).

The equations of motion, applied separately to each region, are simply

$$\nabla \cdot \mathbf{v} = 0 \tag{4.C.6}$$

and

$$\frac{D\mathbf{v}}{Dt} = -\frac{1}{\rho}\nabla P. \tag{4.C.7}$$

We shall solve these equations explicitly in region 1, noting that the solution in region 2 can be obtained from this by letting $\rho_1 \to \rho_2$, $P_1 \to P_2$, and setting $v_0 = 0$. For notational simplicity, we will drop the subscript "2" while solving the equation, and will reintroduce it at the end of the solution.

The equation of continuity becomes

$$\nabla \cdot \mathbf{v} = \nabla \cdot \mathbf{v}_0 + \nabla \cdot \mathbf{v}' = 0,$$

but since at equilibrium,

$$\nabla \cdot \mathbf{v}_0 = 0,$$

we have

$$\nabla \cdot \mathbf{v}' = 0. \tag{4.C.8}$$

Similarly, the Euler equation is

$$\frac{\partial \mathbf{v}_0}{\partial t} + \frac{\partial \mathbf{v}'}{\partial t} + (\mathbf{v}_0 \cdot \nabla)\mathbf{v}_0 + (\mathbf{v}_0 \cdot \nabla)\mathbf{v}' + (\mathbf{v}' \cdot \nabla)\mathbf{v}_0 + (\mathbf{v}' \cdot \nabla)\mathbf{v}' = -\frac{\nabla P_0}{\rho} - \frac{\nabla P'}{\rho}. \tag{4.C.9}$$

This can be considerably simplified by noting that at equilibrium

$$\frac{\partial \mathbf{v}_0}{\partial t} + (\mathbf{v}_0 \cdot \nabla)\mathbf{v}_0 = -\frac{\nabla P_0}{\rho}, \tag{4.C.10}$$

so that

$$\frac{\partial \mathbf{v}'}{\partial t} + (\mathbf{v}_0 \cdot \nabla)\mathbf{v}' + (\mathbf{v}' \cdot \nabla)\mathbf{v}_0 + (\mathbf{v}' \cdot \nabla)\mathbf{v}' = -\frac{\nabla P'}{\rho}. \tag{4.C.11}$$

There are two further simplifications which can be made. First, we note that v_0 is a constant, so that

$$(\mathbf{v}' \cdot \nabla)\mathbf{v}_0 = \left(v'_x \frac{\partial}{\partial x} + v'_y \frac{\partial}{\partial y} + v'_z \frac{\partial}{\partial z} \right)\mathbf{v}_0 = 0.$$

Secondly, we note that we are dealing with a situation in which *small* perturbations to equilibrium are being made. The term $(\mathbf{v}' \cdot \nabla)\mathbf{v}'$ in the above equation is therefore of second order in smallness, while all of the other terms in the equation are of first order. Thus, for small deviations from equilibrium, we can write

$$(\mathbf{v}' \cdot \nabla)\mathbf{v}' \approx 0$$

to give

$$\frac{\partial \mathbf{v}'}{\partial t} + (\mathbf{v}_0 \cdot \nabla)\mathbf{v}' = -\frac{\nabla P'}{\rho}. \tag{4.C.12}$$

We see that the small perturbation approximation leaves us with a linear equation relating the velocity and the pressure, rather than the original nonlinear one. Of course, this equation is much easier to solve than the original one. This technique, which we have used here in the context of a hydrodynamics problem, is called linearization, and is used extensively throughout physics.

If we take the divergence of Eq. (4.C.12) and use the continuity condition that $\nabla \cdot \mathbf{v}' = 0$, we find an equation for the small addition to the equilibrium pressure

$$\nabla^2 P' = 0. \tag{4.C.13}$$

If we substitute the assumed form of P' from Eq. (4.C.5) into this result, we find that

$$-k^2 P'(z) + \frac{\partial^2}{\partial z^2} P'(z) = 0, \tag{4.C.14}$$

which means that the most general solution for $P'(z)$ is just

$$P'(z) = Ae^{-kz} + Be^{kz}, \tag{4.C.15}$$

where A and B are undetermined constants. As in any differential equation, these constants must be determined by the boundary conditions. One boundary condition is that the pressure must stay finite, so that in region 1, we must have $B = 0$, and the perturbation on the pressure must be

$$P'_1 = A e^{-kz} e^{i(kx - \omega t)}. \tag{4.C.16}$$

Similar reasoning in region 2, where z is negative, gives

$$P'_2 = C e^{kz} e^{i(kx - \omega t)}, \tag{4.C.17}$$

where A and C are constants still to be determined.

In order to proceed further, it is necessary to relate the pressure to the displacement of the surface, ξ. We begin by writing down the z-component of the Euler equation [Eq. (4.C.12)] in region 1 (again, dropping the subscript during the derivation), which, with the assumed forms for v' and P' [Eq. (4.C.5)] becomes

$$(-i\omega + ikv_0)v'_z = \frac{k}{\rho} P'. \tag{4.C.18}$$

To relate this to the displacement ξ, we note that $D\xi/Dt$, the velocity of the surface (which is in the z-direction, since ξ is a vector in the z-direction only) must be the same as v'_z, the velocity of a particle at the surface. In other words,

$$\frac{D\xi}{Dt} = \frac{\partial \xi}{\partial t} + v_0 \frac{\partial \xi}{\partial z} = -i\omega\xi + iv_0 k\xi = v'_z \bigg]_{\text{surface}}. \tag{4.C.19}$$

Taking this with Eq. (4.C.18), we find that the pressure in region 1 *at the surface* must be just

$$\frac{k}{\rho_1} P'_1 = (-i\omega + ikv_0)^2 \xi. \tag{4.C.20}$$

A similar argument for region 2 yields

$$-\frac{k}{\rho_2} P'_2 = -\omega^2 \xi. \tag{4.C.21}$$

Now at the surface, we must have

$$P'_1 = P'_2, \tag{4.C.22}$$

so that

$$-\rho_1(\omega - kv_0)^2 = \omega^2 \rho_2, \tag{4.C.23}$$

which can be solved for ω to give

$$\omega = kv_0 \left[\frac{\rho_1 \mp i\sqrt{\rho_1\rho_2}}{\rho_1 + \rho_2} \right],$$
$$= \alpha \mp i\beta. \tag{4.C.24}$$

Thus, the most general form of the time dependence of the quantities P', v', and ξ will be

$$De^{i\alpha t}e^{\beta t} + Ee^{i\alpha t}e^{-\beta t}, \tag{4.C.25}$$

so that for any values of ρ_1 and ρ_2 except the trivial case where $\rho_1 = 0$ or $\rho_2 = 0$, any small perturbation of the surface will be expected to grow with time and the system will be unstable.

Thus, we see that it is indeed possible to determine the stability of a system directly from the equations of motion, simply by assuming small time-dependent deviations from equilibrium, and seeing what sort of time dependence is imposed on the system by the equations and the boundary conditions.

Before leaving this topic, there are a number of points which should be emphasized. First, as was discussed in Chapter 3, the question of stability of a system depends on the type of applied perturbation. It is always possible for a system to be stable against one type of perturbation while being unstable against another.

Second, the fact that we have shown that the time dependence of the perturbation is exponential may at first sight appear unsettling, since such a dependence seems to imply that no matter how small the initial deflections of the surface are, the deviations from equilibrium will approach infinity after a long enough time.

This actually is not the case, as can be seen by examining the velocity v'. If v_0' is the initial perturbation, then at a later time, Eq. (4.C.24) would give

$$v' = v_0'e^{\beta t}.$$

However, in order to derive Eq. (4.C.24), we had to make the linearization hypothesis to get Eq. (4.C.12). Clearly, for large t, this approximation is no longer valid, so that the exponentially growing solution will no longer be valid, either.

The point is that our linearized equations tell us how the system behaves in time near equilibrium, but once the system is far from equilibrium, we have to go back to the original nonlinear equations for a solution. In terms of Fig. 4.12, our results tell us how the ball will roll off

Fig. 4.12. An illustration of a system which behaves differently near equilibrium than it does far from equilibrium.

of the hill, but once we get away from the hill, the situation changes, and we cannot say that the ball will keep rolling forever.

SUMMARY

The velocity field is defined. It is shown that velocity fields which have no curl correspond to fluid motions in which no rotation is present, and velocity fields with zero divergence or symmetric part correspond to motions in which there is no change in densities.

The concept of stability of flow is introduced, and the technique of examining a fluid flow in equilibrium, introducing small, time-dependent perturbations of equilibrium, and applying the equations of motion to the perturbed system is developed. It is argued that if the equations imply that a perturbation, once introduced, grows with time, then it is unstable. This technique is applied to the tangential flow instability problem.

PROBLEMS

4.1. Consider a container on the earth which is filled to a height h with a fluid of density ρ, and has a small opening a distance z down from the top of the fluid, through which a fluid stream can emerge. Assuming irrotational flow, calculate the velocity of the stream just outside the entrance (neglect the effect of the outflow on the height h).

4.2. Consider an imaginary surface Σ inside of a fluid.

(a) Show that the total flow out through the surface is

$$\int_\Sigma \frac{\partial \phi}{\partial n} \, dS,$$

where $\partial / \partial n$ is the derivative normal to the surface.

(b) Use (a) to show that ϕ cannot have a maximum or minimum anywhere inside of the fluid.

(c) Hence show that if the fluid were of infinite extent, and ϕ were not infinite anywhere, we would have to have

$$\phi = \text{const.}$$

everywhere. This is a special case of Liouville Theorem of mathematical analysis.

4.3. Carry out the pictorial analysis given in Section 4.A for the velocity field

$$v_x = Cy^2,$$
$$v_y = Cx^2.$$

4.4. A streamline is defined to be a line which is everywhere tangent to the velocity of the fluid. It can be pictured easily by imagining a small needle inserted into the moving fluid, and a thin stream of dye being emitted from the needle. The dye will mark the fluid in a line which will have the property of a streamline.

Show from the Euler equation that, for general steady-state flow, the quantity

$$\frac{1}{2}\rho v^2 + P + \rho\Omega$$

must be the same everywhere along a given streamline. From your proof, does it follow that the constant in the above expression must be the same for neighboring streamlines? (*Hint*: Write the Euler equation in the form of Eq. (1.B.5), and take the gradient of the equation in the direction of a streamline.)

4.5. Consider a flow of fluid which is in the z-direction, and is axially symmetric, so that

$$v_z = c(r),$$
$$v_\theta = v_r = 0,$$

where $c(r)$ is an arbitrary function. Except for the case $c = 0$, show that it is *not* possible to define a velocity potential for such a flow.

4.6. Why does a flag wave in the breeze?

4.7. Let us reconsider the rings of Saturn problem from the point of view of fluid stability. Consider the rings to be a flat sheet of thickness $2c$, centered on the x-y plane. Let the density of the fluid be ρ, and let the fluid experience a small perturbation such that each plane of the fluid which was level before the perturbation is now displaced by a distance η, where

$$\eta = A\cos mx.$$

(a) Show that the gravitational potential of the perturbed fluid is

$$V_1 = 2\pi\rho cA \ \sin mx e^{-mc}(e^{mz} + e^{-mz})$$

inside the fluid and

$$V_2 = 2\pi\rho cA \sin mxe^{\mp mz}(e^{mc} + e^{-mc})$$

outside.

(b) Calculate the pressure in the fluid to be

$$P = 2\pi\rho(c^2 - z^2) + 2\pi\rho cA \sin mx$$
$$\times [2cm - 1 - e^{-2mc} + e^{-mc}(e^{mz} + e^{-mz})].$$

(c) Hence show that the system is unstable if

$$\lambda_c = \frac{2\pi}{m} > 5.4c.$$

4.8. Consider the two-dimensional flow of an incompressible fluid. Define a *stream function* ψ by the equations

$$v_x = \frac{\partial\psi}{\partial y}, \qquad v_y = -\frac{\partial\psi}{\partial y}.$$

(a) Show that such a definition automatically satisfies the equation of continuity.

(b) Show that for irrotational flow, the equation for the stream function is

$$\nabla^2\psi = 0.$$

(c) Show that the stream function is constant along any streamline.

4.9. We can define a quantity called the *circulation* as

$$\Gamma = \oint_c \mathbf{v} \cdot \mathbf{dl},$$

where the integral is understood to go over any closed path in the fluid. Show that if all of the forces acting on the fluid can be written as the gradient of a potential, that

$$\frac{D\Gamma}{Dt} = 0,$$

i.e. that the circulation is conserved.

4.10. If we define a *complex potential* in terms of the stream function and velocity potential as

$$w = \phi + i\psi,$$

(a) show that w is an analytic function.

(b) Hence (or otherwise) show that the flow of fluid out of an aperture extending into the fluid in a large container (this is called Borda's mouthpiece) will contract half the width of the aperture. (*Hint*: You will want to use complex variable techniques on this problem.)

4.11. Consider two planes meeting at an (acute) angle at the origin. Suppose an incompressible fluid is undergoing potential flow in the corner formed by these planes.

(a) Write down the boundary conditions at the two planes.

(b) Find the solution to the equations of motion and the boundary conditions to lowest power in r, the radial coordinate.

(c) Calculate the velocities of the flow and sketch them out.

(d) Find the streamlines by calculating the stream functions.

4.12. Show that the equation for the velocity potential for the two-dimensional potential flow of an incompressible fluid is

$$\phi = \frac{C}{2\pi} \ln r.$$

(a) Interpret the constant C in terms of the presence or absence of sources of fluid in the system.

(b) Show that if in this case we consider a two-dimensional electrical system, and make the assignments

$$v \to \sigma \mathbf{j}, \quad \phi \to V,$$

where $\mathbf{j}$ is the current density, σ the conductivity and V is the voltage, we get equations which are identical to the hydrodynamic equation.

(c) Hence suggest an experimental method for measuring the flow of a fluid past irregular obstacles.

4.13. Show that the stream function and velocity potential which are due to the motion of a circular cylinder of radius a moving with velocity U parallel to the x-axis are

$$\psi = -\frac{Ua^2 \sin \theta}{r}, \qquad \phi = \frac{Ua^2 \cos \theta}{r}.$$

(*Hint*: Consider a complex potential of the form $w = A/Z$.)

4.14. Consider now a sphere of radius a moving through a fluid with velocity **v**.

(a) Show that the velocity potential (assuming the fluid to be at rest far from the sphere is

$$\phi = \frac{va^3}{2r^2} \cos \theta.$$

(b) Sketch the lines of flow around the sphere.

(c) From the Euler equation, calculate the pressure at the surface of the sphere.

(d) Show that the equation of motion for the sphere in the fluid is just

$$\mathbf{v} = \frac{\sigma - \rho}{\sigma + \frac{1}{2}\rho} \mathbf{F},$$

where **F** is the external force, and σ and ρ are the densities of the sphere and the fluid, respectively. This says that in the absence of an external agent, there is no net force on the sphere. Does this seem reasonable (see Chapter 8)?

REFERENCES

All of the general texts cited in Chapter 1 contain discussions of the velocity potential. The author found the books by Ramsey and Lamb especially readable, and the discussion of tangential instabilities in the Landau and Lifschitz text particularly good.

5

Waves in Fluids

What dreadful noise of waters in mine ears!

WILLIAM SHAKESPEARE
King Richard III, Act I, Scene IV

A. LONG WAVES

One of the most important aspects of the motion of fluids is the wide variety of waves which can be generated and sustained in them. In this chapter, we shall consider three such wave motions, beginning with the long, or tidal, waves in this section. The origin of the name "long waves" will become obvious later in the discussion.

In general, we can think of wave motion as the result of two opposing forces acting on a body. Consider a weight on a spring, for example. If a force is applied which moves the weight away from its equilibrium position, the weight will exert a force which pulls the weight back. If we let go, the spring will return to its equilibrium position, but when it gets there, it will be moving with some velocity. Thus it will overshoot the equilibrium position, and move on until the spring is compressed enough to cause it to reverse its direction. Thus, the existence of the restoring force in the spring leads to the familiar simple harmonic motion.

The situation with fluids is quite similar. Let us consider a body of uniform fluid whose unperturbed height is h (see Fig. 5.1), but whose surface is for some reason perturbed, so that the actual surface is at a height

$$y_s = h + \eta.$$

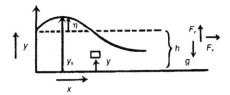

Fig. 5.1.   The perturbed surface of a fluid.

Let us furthermore suppose that this fluid is in a gravitational field on the surface of the earth, so that there is a force ρg per unit volume in the y-direction. Then if $\eta > 0$, the fluid elements in the surface will be pulled downward by gravity, while if $\eta < 0$, the fluid pressure will tend to exert an upward force. Thus, we might expect that we would see harmonic motion in this system.

To make the quantitative ideas introduced in Chapter 4 more definite, we will actually work out the problem mentioned above, with one addition. Let us consider what happens when there is not only a gravitational force acting on the fluid, but an additional force *per unit volume F*, whose components (see Fig. 5.1) are F_x and F_y. We will need these results in Chapter 6 when we discuss the theory of the tides, in which case the extra force would be the gravitational attraction of the moon.

Let us consider an infinitesimal volume element of fluid at a height y in the fluid (see Fig. 5.1). The y-component of the Euler equation is then

$$\rho \frac{Dv_y}{Dt} = -\frac{\partial P}{\partial y} - \rho g + \rho F_y, \tag{5.A.1}$$

while the x-component is

$$\rho \frac{Dv_x}{Dt} = -\frac{\partial P}{\partial x} + \rho F_x. \tag{5.A.2}$$

These equations as they stand are pretty complicated. The most important difficulty is that they are nonlinear. That is, they contain terms in the convective derivative which are proportional to both v and v^2. Such equations are very difficult to solve, and the fact that the Euler equation is nonlinear is the main reason that advances in hydrodynamics are so difficult to make (see Problem 5.2).

To get around this problem, we are going to have to appeal to some of the physics in the problems we are trying to solve. The quantity **v** which appears in the Euler equation refers to the motion of a volume element in a fluid. Now this velocity can be quite small, even though the velocity of the wave in the fluid may be large. This can be seen by thinking about a wave traveling along a rope. Any given segment of the rope moves only a small amount up and down as the wave goes by, but the wave itself may move very quickly. We are going to assume that a similar situation holds in dealing with waves in fluids, and we will write

$$\frac{D\mathbf{v}}{Dt} \approx \frac{\partial \mathbf{v}}{\partial t} \tag{5.A.3}$$

in the Euler equation. This corresponds to saying that since $\mathbf{v}$ is small, we can drop terms of order v^2. It is an approximation which will be made many times in this text. We will see exactly what physical condition is implied by Eq. (5.A.3) later in this section.

If we are dealing with a system like the tides, then the terms F_y in Eq. (5.A.1) which represent the attraction of the moon will be quite small compared to the gravitational force of the earth, so that Eq. (5.A.1) will be given by

$$\rho \frac{\partial v_y}{\partial t} = -\frac{\partial P}{\partial y} - \rho g. \tag{5.A.4}$$

Now if we confine our attention to systems like the tides, there is still another approximation which we can make on this equation. If we think of the tides, we realize that the fluid will move, typically, a distance of several yards in the y-direction over a course of many hours. Thus, the velocity in the y-direction is quite small, and we can expect the rate of change of that velocity to be even smaller. Therefore, it makes sense to set

$$\frac{\partial v_y}{\partial t} \approx 0. \tag{5.A.5}$$

Like the reasoning leading to Eq. (5.A.3), this approximation can be most easily analyzed after we have solved the approximate equations. Physically, this reduces Eq. (5.A.4) to a hydrostatic equation, and amounts to saying that the motion in the y-direction is so slow that we can take it to be such that hydrostatic equilibrium is maintained at all times as far as the y-motion is concerned. This is sometimes called a *quasi-static* approximation. We shall see that this is a valid approximation provided that the depth of the fluid is much less than the wavelength of the wave.

With this final approximation, the left-hand side of Eq. (5.A.4) vanishes, so that the equation can be integrated directly to give

$$P - P_0 = g\rho(h + \eta - y), \tag{5.A.6}$$

where P_0 is the pressure of the medium above the fluid. In most cases, this will just be the atmospheric pressure.

In Eq. (5.A.6), we have a ready incorporated one boundary condition, which is that in this case the pressure must be a constant at $y = h + \eta$. This should be familiar from the discussion of stellar structure in Chapter 2.

We can differentiate Eq. (5.A.6) with respect to x to get

$$\frac{\partial P}{\partial x} = g\rho \frac{\partial \eta}{\partial x}. \tag{5.A.7}$$

The left-hand side of this expression is precisely what appears on the right-hand side of Eq. (5.A.2), so that we can eliminate the pressure between these two equations to get

$$\frac{\partial v_x}{\partial t} = -g \frac{\partial \eta}{\partial x} + F_x. \tag{5.A.8}$$

This equation still contains two unknowns, v_x and η. We can eliminate one of them by recourse to the remaining condition which we can apply to fluids in general, the condition of continuity. We could, of course, simply write it down as in Eq. (1.C.4). However, because we want information about the variable η, we will find it easier to go through the derivation of the equation for the particular geometry in Fig. 5.1.

Consider a wave moving by a point x (see Fig. 5.2), and consider two planes a distance dx apart. The mass of fluid contained between the planes per unit length in the z-direction is just $(h + \eta)\rho \, dx$ so that the time rate of change of mass in the volume is given by

$$\frac{dM}{dt} = \rho \frac{\partial \eta}{\partial t} \, dx. \tag{5.A.9}$$

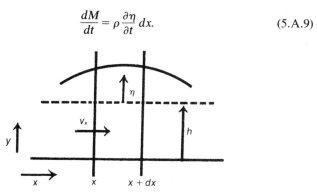

Fig. 5.2. The idea of continuity and the perturbed surface.

How can the mass change? If we are dealing with an incompressible fluid, the only way the amount of mass can change is for some fluid to flow out across the planes. This, in turn, will cause the level of fluid, represented by η, to drop.

The amount of fluid flowing across the left-hand plane is

$$[h + \eta(x)]\rho v_x(x) \approx h\rho v_x(x),$$

where we have dropped the term ηv_x as being second order in small parameters. The amount flowing across the right-hand plane is similarly

$$h\rho v_x(x + dx) \approx h\rho\left[v_x(x) + \frac{\partial v_x}{\partial x}\,dx + \cdots\right],$$

where we have dropped higher-order terms in the Taylor series expansion of v_x. Thus, the net inflow or outflow is the difference between these two quantities, and must be the rate of change of mass in Eq. (5.A.4). Equating these quantities gives

$$-h\frac{\partial v_x}{\partial x} = \frac{\partial \eta}{\partial t} \tag{5.A.10}$$

for the equation of continuity for the incompressible fluid in terms of v_x and η.

If we differentiate Eq. (5.A.10) with respect to t and Eq. (5.A.8) with respect to x, we can eliminate v_x from our equations, and get

$$\frac{\partial^2 \eta}{\partial t^2} = gh\frac{\partial^2 \eta}{\partial x^2} - h\frac{\partial F_x}{\partial x}. \tag{5.A.11}$$

In the case where there is no force except the earth's gravitational field, this becomes

$$\frac{\partial^2 \eta}{\partial t^2} = gh\frac{\partial^2 \eta}{\partial x^2}, \tag{5.A.12}$$

which is simply the wave equation for a wave whose velocity is

$$c = \sqrt{gh}. \tag{5.A.13}$$

The equation has for its solution any function of the traveling wave form, so that

$$\eta(x, t) = f(x - ct), \tag{5.A.14}$$

where f is any wave shape.

Thus, we see that the Euler equation and the equation of continuity lead directly to a wave equation for the deviation of the surface of a fluid from its flat equilibrium configuration when that fluid is under the influence of its own pressure and gravity.

The next question which we must examine is the validity of the approximations which led us to this result. Let us begin with Eq. (5.A.3), which allowed us to drop the nonlinear terms in the Euler equation. Is this approximation really valid?

To examine this question, consider a wave going by a given point in the fluid. Let τ be the time it takes a wave to go past the point, and let η_{max} be the maximum height above h which the surface attains. Then a typical velocity for a particle at the surface would be

$$v_y \sim \frac{\eta_{max}}{\tau}. \tag{5.A.15}$$

It is reasonable to suppose that v_x will be of this order of magnitude as well. Then we expect that the first term in the convective derivative will be roughly

$$\frac{\partial v}{\partial t} \sim \frac{\eta_{max}}{\tau^2}, \tag{5.A.16}$$

since the velocity goes from zero to v in time τ.

By a similar argument, we would have typically

$$(\mathbf{v} \cdot \nabla)\mathbf{v} \sim \frac{1}{\lambda}\left(\frac{\eta_{max}}{\tau}\right)^2, \tag{5.A.17}$$

where λ is the wavelength of the wave. Thus, we will have

$$(\mathbf{v} \cdot \nabla)\mathbf{v} \ll \frac{\partial \mathbf{v}}{\partial t},$$

provided that

$$\eta_{max} \ll \lambda. \tag{5.A.18}$$

In other words, whenever the wavelength of the wave is long compared to typical distances which particles in the fluid move while the wave goes by, we can drop the term in $(\mathbf{v} \cdot \nabla)\mathbf{v}$. Since this condition is easily met by most waves, we shall not refer to this approximation again, but we will use it throughout the remainder of the discussion.

The second important approximation was stated in Eq. (5.A.5), where we assumed that the y-equation could be treated in the quasi-static limit. To examine this approximation, we note that had we not used Eq. (5.A.5), we would have to replace Eq. (5.A.6) by

$$P - P_0 = g\rho(h + \eta - y) + \rho \int_y^{h+\eta} \frac{\partial v_y}{\partial t} \, dy, \tag{5.A.19}$$

where the last term represents the effect of v_y. Now if we denote the maximum acceleration of a particle in the y-direction by β, then we have

$$\rho \int_y^{h+\eta} \frac{\partial v_y}{\partial t} \, dy \leq \beta \rho (h + \eta - y) \leq \beta \rho (h + \eta),$$

where the second inequality follows from the fact that the expression $(h + \eta - y)$ has its largest possible value at $y = 0$.

On the other hand, the first term on the left-hand side of Eq. (5.A.19) will have its minimum near the surface, and its minimum value will be of order $g\rho\eta$. Thus, we can always drop the correction term provided that

$$\beta \ll \frac{g\eta}{h}, \tag{5.A.20}$$

since in that case, the maximum correction is less than the minimum of the term to which it is being compared.

We can now proceed using the same type of arguments that were used before. If the typical acceleration is

$$\beta \sim \frac{\eta}{\tau^2},$$

where, by definition, $\tau = \lambda / c$, then Eq. 5.A.20 becomes

$$\frac{c^2}{\lambda^2} \ll \frac{g}{h},$$

which, using Eq. (5.A.13), we can finally write as

$$\lambda \gg h. \tag{5.A.21}$$

Thus, the quasi-static approximation is valid provided that the wavelength of the waves in question are much greater than the depth of the fluid. There are many examples of such cases (some of which are given in the problems at the end of the chapter). For example, if we were dealing with tides, this would clearly be a valid approximation, since the length of the tidal bulge is on the order of the circumference of the earth, while the depth of the ocean is only a few kilometers. Another example would be waves approaching a beach, since at some point the depth of the fluid will become small enough to satisfy Eq. (5.A.21).

B. SURFACE WAVES IN FLUIDS

In the previous section, we saw that if we made a series of approximations on the Euler equations and the equations of continuity, we could derive a wave equation for η, the displacement of the surface from equilibrium. The most important assumption was the long-wave approximation, which, in the form of Eq. (5.A.3) allows us to neglect the vertical motion of the fluid elements. This assumption is valid in many cases of interest, but it is clear that there are many cases where it is not.

If we were to consider waves on the ocean or a lake, for example, the long-wave length approximation would not apply. Such waves typically have wavelengths of the order of tens or hundreds of feet, which is much less than the depth of the water. This means that we will have to go through the derivation without the benefit of Eq. (5.A.3).

For problems of this type, it is very convenient to use the velocity potential defined in Section 4.B. Let us assume that we are dealing with irrotational flow of an incompressible fluid, so that the equation for the potential is

$$\nabla^2 \phi = 0. \tag{5.B.1}$$

The use of this equation already incorporates the equation of continuity, so that the only other equation which we need to write down is the Euler equation. In terms of the velocity potential, this is given in Eq. (4.B.5). If we make the usual assumption that we can drop second-order terms in the velocity, this is just

$$\frac{P}{\rho} + \frac{\partial \phi}{\partial t} + \Omega = 0. \tag{5.B.2}$$

However, the role which this equation will now play is somewhat different than in the previous section. There, we combined the Euler equation and the equation of continuity to display the wave equation explicitly. In this section, we shall proceed by *assuming* that ϕ has a wave-like solution, and verify that this is indeed the case by direct substitution into the above two equations. We shall see that in this case, the Euler equation enters only in that it determines the boundary conditions at the fluid surface.

Let us, then, guess that it is possible to find solutions of Eqs. (5.B.1) and (5.B.2) of the form

$$\phi = f(y) \cos (kx - \omega t), \tag{5.B.3}$$

where $f(y)$ is some function to be determined. Putting this into the Laplace equation gives

$$\frac{d^2 f(y)}{dy^2} - k^2 f(y) = 0, \tag{5.B.4}$$

which means that the most general form of $f(y)$ is

$$f(y) = Ae^{ky} + Be^{-ky}. \qquad (5.B.5)$$

To proceed further, we need to impose the boundary conditions. At the bottom of the fluid, at $y = 0$, we know that $v_y = \partial\phi/\partial y = 0$, since, by definition, no fluid can cross the bottom boundary. This means that

$$\left.\frac{\partial f}{\partial y}\right]_{y=0} = 0,$$

so $A = -B$, and

$$f(y) = 2A \cosh(ky). \qquad (5.B.6)$$

The second boundary condition is just as simple physically, but somewhat more difficult mathematically. It states that at the surface of the fluid, the pressure must be equal to P_0, the atmospheric pressure, which we take to be a constant. To state this condition at the perturbed surface, we shall have to make use of the Euler equation in the form (5.B.2).

At the surface of the earth, we can take

$$\Omega = gy + \text{const.}, \qquad (5.B.7)$$

so that at $y = h + \eta$, Eq. (5.B.2) becomes

$$\frac{P_0}{\rho} + \frac{\partial\phi}{\partial t}\bigg)_{y=h+\eta} + gh + g\eta = \text{const.}$$

Now the second term on the left can be expanded

$$\frac{\partial\phi}{\partial t}\bigg)_{y=h+\eta} = \frac{\partial\phi}{\partial t}\bigg)_{y=h} + \frac{\partial^2\phi}{\partial t\,\partial y}\bigg)_{y=h} \eta + \cdots,$$

and if we drop all but the first term in the expansion as being of second order in small quantities, and at the same time define the constant in Eq. (5.B.7) appropriately, we find

$$-\eta = \frac{1}{g}\frac{\partial\phi}{\partial t}\bigg)_{y=h}, \qquad (5.B.8)$$

or, differentiating

$$-\frac{\partial\eta}{\partial t} = \frac{1}{g}\frac{\partial^2\phi}{\partial t^2}\bigg]_{y=h}. \qquad (5.B.9)$$

Now if we consider a volume element just at the surface of the fluid, it has a velocity given by $v_y = \partial\phi/\partial y)_{y=h+\eta}$. But the volume element at the surface must be moving with just the velocity itself, which is $+\partial\eta/\partial t$.

Thus, we have at the surface

$$v_y = \frac{\partial \eta}{\partial t},$$ (5.B.10)

or

$$\left[\frac{\partial^2 \phi}{\partial t^2} + g \frac{\partial \phi}{\partial y} \right]_{y=h} = 0,$$ (5.B.11)

where, as in the derivation of Eq. (5.B.8), we have replaced all quantities which are to be evaluated at the surface $y = h + \eta$ by quantities evaluated at the surface of equilibrium $y = h$.

Equation (5.B.11), then, is the boundary condition at the upper surface (the analogue of $v_y(0) = 0$, the boundary condition at the lower surface) which our assumed solution has to satisfy. Inserting the solution in Eqs. (5.B.8) and (5.B.6), we find that we can satisfy Eq. (5.B.11) provided that

$$\omega^2 = gk \tanh (kh).$$ (5.B.12)

Thus, the general solution for the velocity potential is just

$$\phi = 2A \cosh (ky) \cos (kx - \omega t)$$ (5.B.13)

and, since the solutions to Laplace's equation are unique, this is the only solution. To find the surface displacement, we use the boundary condition Eq. (5.B.8) to get

$$\eta = -\frac{2\omega A}{g} \cosh (ky) \sin (kx - \omega t) \Big]_{y=h},$$ (5.B.14)

which describes a wave traveling in the x-direction, as we expected. The velocity is given by

$$c = \frac{\omega}{k} = \left(\frac{g}{k} \tanh kh \right)^{1/2} = \left(\frac{g\lambda}{2\pi} \tanh \left(\frac{2\pi h}{\lambda} \right) \right)^{1/2},$$ (5.B.15)

where we have written $k = 2\pi/\lambda$.

Recalling that

$$\tanh x = \begin{cases} 1 & x \gg 1 \\ x & x \ll 1, \end{cases}$$

we find

$$c = \begin{cases} \sqrt{gh} & \lambda \gg h \\ \sqrt{\dfrac{g\lambda}{2\pi}} & \lambda \ll h. \end{cases}$$ (5.B.16)

Thus, in the long wavelength limit, we recover the long-wave result which we derived in the last section. However, when the depth of the fluid is comparable to or shorter than the wavelength, we find that the velocity depends on the wavelength itself, which is a result which we have not encountered before.

A question which we might well ask at this stage is why the relation of the depth to the wavelength of the wave should be important. To answer this question, let us calculate the velocities of volume elements in the fluid at some depth y [this velocity is not to be confused with the velocity of the wave, which is given by Eq. (5.B.15)]. From the definition of the velocity potential,

$$v_y = 2kA \sinh(ky) \cos(kx - \omega t),$$
$$v_x = -2kA \cosh(ky) \sin(kx - \omega t), \qquad (5.B.17)$$

at an arbitrary point in the fluid. Thus, each particle is seen to describe an ellipse in the x-y plane, with the axis in the y-direction being proportional to $\sinh ky$, and the axis in the x-direction to $\cosh ky$ (see Problem 5.3).

There are several conclusions which can be drawn from this. First, at $y = 0$, the vertical movement vanishes (this was to be expected, since it was our first boundary condition). More important, we see that the disturbance associated with the wave falls off like a hyperbolic function as we go below the surface, and the length associated with this fall off is $1/k$, or $\lambda/2\pi$. Thus, the disturbance is confined to something like a distance of one wavelength from the surface. This is the origin of the name "surface wave" and of the dependence of the solution of the equations of fluid mechanics on the relation between depth and wavelength. One could say that the existence of a wave requires the cooperation of the fluid at the surface to a depth about equal to the wavelength of the wave. In the long wavelength limit, this means that we must have the entire fluid involved in the wave.

In Section 12.E, we shall see that this surface wave phenomenon is not unique to fluids, but exists in solids as well.

C. SURFACE TENSION AND CAPILLARY WAVES

Up to this point, we have considered only pressure and external forces acting on particles of the fluid. While this may be a perfectly adequate description in the interior of the fluid, it is well known that there are forces on the surface of a fluid which tend to oppose any increase in surface area—any "stretching" of the surface. This force is usually called the "surface tension," T, and is defined by the work necessary to increase

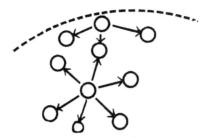

Fig. 5.3.   Molecular forces and surface tension.

the area of a surface by an amount dS by the relation

$$dW = T\,dS. \tag{5.C.1}$$

The simplest way to picture the reason for this force is to note (see Fig. 5.3) that there are usually attractive (cohesive) forces on the molecular level in a fluid which tends to make it stay together. For a molecule in the interior, these forces are exerted in all directions, and therefore cancel out on the average. For a molecule on the surface, however, these forces are all directed inward toward the body of the fluid, and there is a net inward force. Increasing the surface area corresponds to putting more particles into the surface, and hence work must be done against the attractive forces, giving rise to relation (5.C.1) above. We should note that in terms of this picture, the existence of surface tension is strictly a geometrical effect—it arises because a surface, by definition, divides a region filled with fluid from a region empty of the fluid. Thus, whether the force is molecular in origin (as in the section) or is a consequence of nuclear interactions (as is the case of the liquid drop model of the nucleus which we shall discuss later) will not affect the existence of a surface force.

In order to quantify the above remarks on surface tension, let us examine the following problem: A surface finds itself with pressure P_1 on one side and P_2 on the other. The imbalance of pressures causes the surface to expand. In the process, work must be done against T, the surface tension. Consider the element to have unperturbed lengths δl_1 and δl_2 and radii of curvature R_1 and R_2 (see Fig. 5.4) and let the lengths of the sides after stretching be given by $\delta l_1(1 + \alpha)$ and $\delta l_2(1 + \beta)$. Then, since

$$\alpha = \frac{dx}{R_1}, \qquad \beta = \frac{dx}{R_2},$$

we have

$$W = T\,dA = T\,dl_1\,dl_2(\alpha + \beta).$$

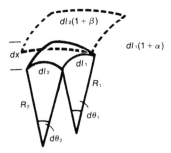

Fig. 5.4.   The displacement of a surface by pressure differentials.

On the other hand, the work done by the pressure in displacing the surface a distance dx is just

$$(P_2 - P_1)\, dl_1\, dl_2\, dx,$$

so that

$$P_2 - P_1 = T\left(\frac{1}{R_1} + \frac{1}{R_2}\right). \tag{5.C.2}$$

Thus, we find that the surface force is quite large when the surface is sharply curved. This new force introduces a rather different problem at the surface. Up to this point, we have always used the condition that a surface was characterized by a constant value of the pressure. But the existence of a force in the surface which could balance a force due to an imbalance in pressure means that we must be more careful. Equation (5.C.2) now tells us that the condition at the surface is no longer that P is constant, but that pressure differences are related to the curvature of the surface, and that changes in curvature along the surface will necessitate changes in the pressure difference across it.

Turning our attention now to the effect of surface tension on the type of waves we have been discussing, let us consider the situation shown in Fig. 5.5, where two semi-infinite fluids of densities ρ and ρ' have an interface at the plane $y = 0$. If we let ϕ and ϕ' represent the velocity potentials in the two fluids, then, as before, the basic equations governing the

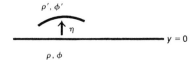

Fig. 5.5.   The perturbed interface between two fluids.

potentials are

$$\nabla^2 \phi = 0,$$
$$\nabla^2 \phi' = 0,$$

(5.C.3)

and, following the steps in the previous section, we find

$$\phi' = (C' e^{ky} + C e^{-ky}) \cos(kx - \omega t),$$
$$\phi = (D e^{ky} + D' e^{-ky}) \cos(kx - \omega t).$$

(5.C.4)

As usual, we will determine the constants C and C', D and D' from the boundary conditions. From the requirement that the velocities stay finite at $y = \pm \infty$, we find

$$C' = D' = 0.$$

(5.C.5)

If, as before, we denote by η the deviation of the surface from equilibrium, and we assume, following the procedure of the previous section, that

$$\eta = A \sin(kx - \omega t),$$

(5.C.6)

then the condition that an element in the surface move at the same velocity as the surface itself gives

$$\frac{\partial \eta}{\partial t} = \frac{\partial \phi}{\partial y}\bigg)_{y=0} = \frac{\partial \phi'}{\partial y}\bigg)_{y=0},$$

(5.C.7)

which yields

$$D = -\frac{\omega}{k} A,$$
$$C = \frac{\omega}{k} A.$$

(5.C.8)

With these solutions for the velocity potentials, we can now solve the Euler equation for the pressure on each side of the surface. We find

$$\frac{P}{\rho} = \left(-\frac{\omega^2}{k} - g\right) A \sin(kx - \omega t)$$

(5.C.9)

and

$$\frac{P'}{\rho'} = \left(\frac{\omega^2}{k} - g\right) A \sin(kx - \omega t).$$

(5.C.10)

By Eq. (5.C.2), this is supposed to be related to the surface tension and the curvature of the surface. From Fig. 5.5, we see that the radius of curvature in the z-direction is just

$$\frac{1}{R_2} = 0,$$

(5.C.11)

since by hypothesis, nothing depends on the z-coordinate. From Problem 5.7, or from elementary calculus, we know that the other radius is given by

$$\frac{1}{R_1} = \frac{\partial^2 \eta / \partial x^2}{[1 + (\partial \eta / \partial y)^2]^{1/2}} \approx \frac{\partial^2 \eta}{\partial x^2}, \qquad (5.C.12)$$

where the second approximate equality is true for small deformations of the surface. Substituting Eqs. (5.C.9) and (5.C.10) into Eq. (5.C.2) gives, after some cancellation, the condition that

$$\omega^2 = gk \frac{(\rho' - \rho)}{\rho + \rho'} + \frac{Tk^3}{\rho + \rho'}. \qquad (5.C.13)$$

If we recall that the velocity of the wave, as opposed to velocity of the fluid particles, is given by

$$c = \frac{\omega}{k}, \qquad (5.C.14)$$

we see that

$$\begin{aligned}
c^2 &= \frac{\rho' - \rho}{\rho + \rho'} \frac{g}{k} + \frac{T}{\rho + \rho'} k, \\
&= \frac{\rho' - \rho}{\rho + \rho'} \frac{g\lambda}{2\pi} + \frac{T}{\rho + \rho'} \frac{2\pi}{\lambda},
\end{aligned} \qquad (5.C.15)$$

where the second equality follows from the definition $k - 2\pi/\lambda$.

There are a number of interesting consequences of this result. We see that if we take the limit

$$\rho' = T = 0.$$

we get

$$c^2 = \frac{g}{k}, \qquad (5.C.16)$$

which is precisely the result for surface waves in a fluid of infinite depth [see Eq. (5.B.16)]. This gives us some confidence that our results are correct, since our intuition tells us that the problem we are working in this section should reduce to the problem of the previous section in this limit.

A related consequence comes if we note that for very large wavelengths, the second term in Eq. (5.C.15) will become unimportant, and the wave will look like an ordinary surface wave, regardless of the presence of surface tension. On the other hand, at very small wavelengths, the second term will dominate completely, and we will have

$$c = \left(\frac{2\pi T}{\rho + \rho'} \frac{1}{\lambda} \right)^{1/2}. \qquad (5.C.17)$$

A pictorial way of representing this is to plot c^2 versus wavelength (see Fig. 5.6).

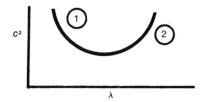

Fig. 5.6. A plot of velocity versus wavelength. The small wavelength part corresponds to capillary waves, and the long wavelength part to surface waves.

In region 2, we have the ordinary surface waves discussed in the previous section. For such waves, the existence of surface tension is largely irrelevant. In region 1, we have a new type of wave, whose existence is a direct consequence of the existence of surface forces. This type of wave is generally called a *capillary wave*, or *ripple*.

The reader has probably already observed capillary waves in nature. When a wind is blowing on a lake or the ocean, one often sees the usual large waves, but with small ruffles superimposed on them. The ruffles are, in fact, capillary waves which are caused by the wind (see Problem 5.16).

We will see other examples of surface tension effects in Chapter 8, when we discuss nuclear fission, and in Chapter 14, when we discuss some applications to medicine.

SUMMARY

We have seen that as a consequence of the Euler equation and the equation of continuity that there are a wide variety of waves possible in fluids. These include long waves, in which the vertical motion of the fluid can be ignored so long as the depth of the fluid is much less than the wavelength; surface waves, in which the disturbance of the wave diminishes with depth in the fluid, and capillary waves, which depend on the existence of surface tension, and are typically of short wavelength. This does not exhaust the number of possible waves in fluids, but represents the types of waves most commonly encountered in physical situations.

PROBLEMS

5.1. Show that there are wavelike disturbances (for long waves) possible on a canal of rectangular cross section and uniform depth, where the frequency of the wave is

$$\omega = \frac{n\pi}{l}\sqrt{gh},$$

where l is the length and h the depth of the canal, and n is an integer. An oscillation of this type on a surface, which can be excited by earthquakes, for example, is called a seiche, and is similar to the phenomenon of water sloshing around in a bathtub.

Using a reasonable approximation scheme, calculate the period of a seiche in (1) Lake Geneva, Switzerland, and (2) Lake Eyre, Australia.

5.2. In Section 5.A, we discussed nonlinear equations briefly. To see why such equations are difficult to solve, consider the equation

$$\alpha \frac{df}{dx} + \beta \left(\frac{df}{dx}\right)^2 = g(x).$$

(a) Show that if f_1 and f_2 are solutions to this equation,

$$f = f_1 + f_2$$

is not necessarily a solution.

(b) Show that if $f_1(x)$ is a solution,

$$f = Cf_1(x),$$

where C is a constant, is not necessarily a solution.

(c) Could we solve such an equation by solving for one Fourier component, and then adding components together?

5.3. Given the equations for the velocity of fluid elements of a surface wave in Eq. (5.B.17), show that the motion described by a fluid element is indeed the ellipse described in the text.

5.4. In many parts of this text, we shall use the incompressible fluid approximation. That is, we shall write the equation of continuity as

$$\nabla \cdot \mathbf{v} = 0.$$

The physical reason for this is based on the fact that the volume of most fluids is relatively insensitive to changes in pressure. Convince yourself that this is true by looking at several different fluids, including water.

5.5. (a) Show that in the case of a canal in which the breadth b and depth h vary along the length of the canal, the equation for long waves becomes

$$\frac{\partial^2 \eta}{\partial t^2} = \frac{g}{b(x)} \frac{\partial}{\partial x} \left(h(x)b(x) \frac{\partial \eta}{\partial x}\right).$$

We can make a simple model of a tidal inlet, or river estuary, as a system in which the depth varies uniformly from h_0 at the ocean to zero at a distance a from the ocean, and whose breadth varies from b_0 to zero over the same range. Show that if the elevation at the ocean is given by

$$\eta = A \cos \sigma t,$$

the elevation in the estuary is given by

$$\eta = A\frac{J_1(2\sqrt{kx})}{\sqrt{kx}}\cos \sigma t,$$

where x is the coordinate measuring distance from the ocean, and

$$k = \frac{\sigma^2 a}{gh_0}.$$

(b) Show that if the breadth is constant, but the depth varies as above, the elevation varies as

$$\eta = AJ_0(2\sqrt{kx})\cos \sigma t.$$

(c) Consider the sloping bottom of a beach as the canal of variable depth in this problem. Suppose that 10-foot breakers are coming in off the ocean at the rate of one every 10 seconds. What would the slope of the beach have to be so that (1) the long-wave solution is valid, and (2) the surf near the beach is at least three feet high?

5.6. In Sections 5.B and 5.C, we assumed a form for the velocity potential and the surface disturbance of the form $\cos (kx - \omega t)$ or $\sin (kx - \omega t)$. Another commonly used form would be $e^{i(kx-\omega t)}$. Show that the final results in Eqs. (5.B.15) and (5.C.15) are unchanged if we use this exponential form.

5.7. Show that the radius of curvature of a curve $y(x)$ is given by

$$\frac{1}{R} = \frac{d^2y/dx^2}{(1 + (dy/dx)^2)^{1/2}}.$$

5.8. Consider the case of the type shown in Fig. 5.5, in which the upper region is a vacuum, so that $\rho' = 0$, while the surface tension of the lower fluid is T. Suppose also that the depth of the lower fluid is h. Derive an expression for the velocity of the wave in this case, and show that it reduces to Eq. (5.C.15) in the limit $T \to 0$.

5.9. Equation (5.B.10) can be derived in another way. Consider an element of surface dS which will move an amount $\Delta\eta$ in the vertical direction in time Δt. From the conservation of mass, show that

$$\frac{\Delta\eta}{\Delta t} = v_y.$$

5.10. An important type of wave which can propagate in a fluid is the *sound wave*. Unlike the waves considered in the text, these waves can exist only in compressible fluids.

(a) Assume that the density of a fluid is given by

$$\rho = \rho_0(1 + s),$$

where $s \ll 1$. Show that if the velocities are small, the equation of continuity is

$$\frac{\partial s}{\partial t} = \nabla^2\phi.$$

(b) From the Euler equation, show that

$$\frac{\partial^2 \phi}{\partial t^2} = c^2 \nabla^2 \phi,$$

where $c^2 = \partial p / \partial \rho)_0$.

(c) Show that a wave of the type derived in part (b) is, in fact, a wave in which the density of the fluid is changing periodically with time. Hence show that c must be the velocity of sound in the fluid.

5.11. Show that the only sound waves that can exist in a closed tube of length L are those for which the displacement of the particles at a point x is given by

$$\xi \propto \sin\left(\frac{n\pi x}{L}\right) \cos\left(\frac{n\pi ct}{L}\right).$$

Discuss the construction of an organ pipe.

5.12. Suppose that there are two media, separated by the plane $x = 0$. Suppose further that the velocity potential, density, and velocity of sound in the first medium are ϕ_1, ρ_1, and c_1 with similar definitions for the second medium.

(a) Write the equations governing ϕ and s in each medium, and the boundary conditions which can be expected to hold at the interface (see Problem 5.10).

(b) Suppose that a plane wave of frequency ω is incident at an angle θ to the normal from the upper medium. If θ_1 is the angle of the refracted wave, show that

$$\frac{\sin \theta}{c} = \frac{\sin \theta_1}{c_1}.$$

(c) If A, A', and A_1 are the amplitudes of the incident, reflected, and refracted waves at the interface, show that

$$\frac{A}{\dfrac{\rho_1}{\rho} + \dfrac{\cot \theta_1}{\cot \theta}} = \frac{A'}{\dfrac{\rho_1}{\rho} \dfrac{\cot \theta_1}{\cot \theta}} = \frac{A_1}{2}.$$

(d) When will there be no reflected wave?

5.13. Write down the equation which governs the propagation of a sound wave in a spherically symmetric uniform medium (see Problem 5.10).

(a) Show that the equations for ϕ and s yield, at large r, a wave for which $v = cs$. Show that this same equation holds true for plane waves.

(b) Show that if a source at $r = 0$ causes a velocity potential which varies as

$$\phi \sim e^{i\omega t},$$

that the velocity potential for an outgoing wave will have the form

$$\phi = \frac{A}{4\pi r} \cos \omega\left(t - \frac{r}{c}\right).$$

(c) Show that mean work done by the source is

$$Q = \frac{\rho \omega^2 A^2}{8\pi c}.$$

5.14. Consider a fluid of density ρ and surface tension T in a box of depth h with a flexible bottom. Suppose that the bottom is manipulated so that its displacement from a plane is given by

$$\eta = A \cos (\omega t - kx).$$

Show that the surface of the fluid will be given by

$$y = A' \cos (\omega t - kx),$$

where

$$A' = A \left[\cosh (kh) - k \left(\frac{g + k^2 T/\rho}{\omega^2} \right) \sinh (kh) \right]^{-1}.$$

5.15. You have probably had the experience of walking somewhere with a cupful of coffee and have observed the standing waves which can be set up in such a system.

(a) If the cup is of circular cross section, radius a, and of depth h, show that the general standing wave on it is of the form

$$\eta \sim A J_n (kr) \cos n\theta \cos \sigma t.$$

(b) Determine the values of k which satisfy the boundary conditions.

(c) Determine the frequency of oscillation of the waves, given the known velocity of long waves $c = \sqrt{gh}$.

(d) How would you prevent the coffee from spilling over?

5.16. Retrace the development in Section 5.C for the case in which the upper medium is moving with velocity U with respect to the lower medium.

(a) Show the Eq. (5.C.15) is now replaced by

$$c = \frac{U\rho'}{\rho + \rho'} \pm \left[\frac{\rho - \rho'}{\rho + \rho'} \frac{g}{k} + \frac{kT}{\rho + \rho'} - \frac{U^2 \rho \rho'}{(\rho + \rho')^2} \right]^{1/2}.$$

(b) This is clearly a model for waves generated by the wind. Can we ever get a situation in which waves travel *against* the wind? Interpret this result.

(c) What is the value of U for which the perturbation at the surface will be unstable? Show that for water, $U \gtrsim 6.5$ m/sec will cause the waves to be blown into spindrift.

5.17. Show that a fluid in space (with no gravitational field around) will form itself into a sphere. Hence comment on the prospects of manufacturing ball bearings in sattelites.

REFERENCES

All of the general texts cited in Chapter 1 contain discussions of fluid waves. In particular, the text by Lamb, in Chapters 8, 9, and 10, contains a large number of physically interesting examples of wave motion, including the ship's wake and tidal waves.

6

The Theory of the Tides

A ring from his finger he hastily drew
Saying, "Take it, dearest Nellie, that your heart may be true.
For the good ship stands waiting for the next flowing tide
And if ever I return again, I will make you my bride."

Traditional English Ballad

A. THE TIDAL FORCES

The tides have always played an important role in human affairs. In the last chapter, we showed that the equations governing the motion of fluids admit wavelike solutions, but we did not address ourselves to the question of how such motions might be generated. In this chapter, we will look at one type of wave—the long wave—and show how the waves are generated and how they might be expected to behave in some simple models of the oceans.

It is generally known that the tides are caused by the effects of the moon's gravitational attraction on the water in the oceans. Let us begin our consideration of the theory of the tides by working out an approximate expression for the potential which describes this attraction. Consider the geometry shown in Fig. 6.1. The gravitational potential at the point P due to the moon is just

$$\Omega_M(r) = \frac{-GM_m}{[r^2 + D^2 - 2rD \cos \Theta]^{1/2}}, \qquad (6.A.1)$$

where r is the distance from the center of the earth to P. However, this is not the potential which we would have to use if we wish to calculate the tides. The reason for this is that in addition to exerting a force on the water at the earth's surface, the moon also accelerates the earth as a whole. It is

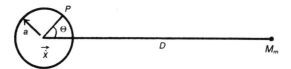

Fig. 6.1. The configuration of the earth and the moon.

only the net acceleration, of course, which would be measured by an observer at the surface of the earth. The acceleration of the center of the earth because of the presence of the moon is

$$\frac{GM_m}{D^2}\,\hat{x} = \frac{\partial}{\partial x}\left[\frac{GM_m}{D^2}\,x\right]\hat{x} = \nabla\left(\frac{GM_m}{D^2}\,x\right), \qquad (6.A.2)$$

where we have written the force as the derivative of a function, which we can now regard as a potential, that takes into account the motion of the earth, and $\hat{x}$ is a unit vector in the x-direction. Thus, the net gravitational potential at P—the net potential which will actually be felt by the water—is just

$$\Omega_D = \frac{-GM}{(D^2 + r^2 - 2rD\cos\Theta)^{1/2}} + \frac{GM}{D^2}\,r\cos\Theta. \qquad (6.A.3)$$

We have written this as Ω_D, the disturbing potential, to distinguish it from Ω_M, the potential at P due to the moon. Now in practice, we know that $r/D \ll 1$, so we can expand Ω_D to lowest order in r/D to get

$$\Omega_D = \frac{GMa^2}{2D^3}(1 - 3\cos^2\Theta),$$

$$= H_D\left(\cos^2\Theta - \frac{1}{3}\right), \qquad (6.A.4)$$

where we have let $r = a$ in the final step, and thus restricted our attention to the surface of the earth. We have also set the zero of Ω_D at $-GM/D$. It is this potential whose derivatives are the "extra" forces which were introduced in Eqs. (5.A.1) and (5.A.2). In fact, we have

$$F_x = -\frac{\partial}{\partial x}\,\Omega_D. \qquad (6.A.5)$$

B. TIDES AT THE EQUATOR

As a first example of a theory of the tides, let us consider a case in which the geometry is as simple as possible, so that we can see the physics of the situation clearly. Let us consider an observer at the equator, and let us assume that the moon lies directly above the equator at all times. Let us furthermore neglect the dynamical effects of the earth's rotation (i.e. neglect centrifugal and Coriolis forces), and let the only effect of this rotation be an apparent movement of the moon (as seen by our observer) around the earth once each day. We will also assume that the earth is a uniform sphere covered with an ocean of uniform depth, and ignore the presence of land masses.

In this case, the angle Θ which appears in Eq. (6.A.4)—the angle between the vector to the point P at which the tides are being measured and the vector to the moon—will lie in the plane of the equator. This greatly simplifies the geometry, since the angle Θ now corresponds to the angle of longitude at the equator (see Fig. 6.2). The complications which arise with the more general case will be discussed in the next sections. We shall see, in fact, that the main mathematical complications which appear in the Laplace theory of the tides have to do with the fact that the angle Θ between the radius to the point of observation and the radius to the moon is not, in general, so easily expressible in terms of other angles in the problem.

In deriving the long-wave equation, Eq. (5.A.11), we used Cartesian coordinates. For an observer on the surface of the earth, the apparent vertical and horizontal would be the x- and y-axes shown in Fig. 6.2. Since it is only the x-component of the extra force which enters Eq. (5.A.11), we have

$$F_x = -\frac{\partial}{\partial x}\,\Omega_D = -\frac{1}{a}\,\frac{\partial}{\partial \phi}\left[\frac{1}{2}\frac{GMa^2}{D^3}\,(1 - 3\cos^2\phi)\right]$$

$$= \frac{3}{2}\frac{GMa}{D^3}\sin 2\phi, \qquad (6.B.1)$$

where we have used the geometrical identity $dx = a\,d\phi$ and set $\Theta = \phi$. If we insert Eq. (6.B.1) into Eq. (5.A.11), we find

$$\frac{\partial^2\eta}{\partial t^2} = \frac{gh}{a^2}\frac{\partial^2\eta}{\partial \phi^2} + \frac{3GMh}{D^3}\cos 2\phi. \qquad (6.B.2)$$

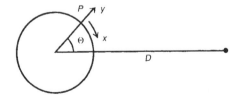

Fig. 6.2.   The coordinates for the discussion of long waves.

From the theory of inhomogeneous differential equations (see Appendix E), we know that the most general solution of Eq. (6.B.2) can be written

$$\eta = \eta_h + \eta_P, \tag{6.B.3}$$

where η_h represents the solution to the equation with $F_x = 0$ (the homogeneous solution) and η_P represents the particular solution for the equation with the forcing term.

In what follows we shall not include the term η_h in our solutions, but look only for the particular solutions to the equations. The reason for this lies with our physical intuition, and not with the mathematics. We know that we have ignored processes (such as friction and viscosity) by which a real fluid will lose energy. We know, therefore, that a disturbance in the fluid will tend to die out unless some outside agency is present which adds energy continuously to the system. In the case we are considering, this outside agency is, of course, the moon. Thus, we know that the only long-term disturbances which will be present in the oceans will be those represented by η_P, while the disturbances represented by η_h will tend to die out with time. It should be noted that this same sort of treatment of long- and short-term effects is often encountered in electrical circuits, where the homogeneous solutions are customarily referred to as transients, and the particular solutions are referred to as steady-state solutions.

If we let ω be the frequency of the moon about the earth as observed from the point P, then

$$\phi = \omega t, \tag{6.B.4}$$

where we set the zero of time when the moon is directly over the point P. ω, of course, should correspond to a period of 24 hours. Equation (6.B.4) means that we can eliminate the variable ϕ from Eq. (6.B.2) and get

$$\frac{\partial^2 \eta}{\partial t^2} = \frac{3GMah}{D^3 \left(1 - \dfrac{gh}{\omega^2 a^2}\right)} \cos^2 \omega t, \tag{6.B.5}$$

which is easily solved to give

$$\eta = \frac{3GMa^2 h}{4D^3(gh - \omega^2 a^2)} \cos^2 \omega t. \tag{6.B.6}$$

There are two important features of this solution of the tidal equations at the equator which we should note. First, we observe that the water level at a particular point will reach its maximum value twice a day—even though the moon traverses its path only once in the same period of time.

This feature of the tides—that they are semi-diurnal—will reappear when we discuss the Laplace theory later.

Perhaps more interesting is the fact that if we look at $t = 0$, the time when the moon is directly overhead, η will attain either its maximum or minimum value, depending on the sign of $gh - \omega^2 a^2$. We recall that $c^2 = gh$ is the velocity of a long wave, and we see that ωa is the velocity of the moon's "shadow" on the earth. If we take the earth to have a radius of 4000 km, and the average depth of the ocean as 4 km, we easily see that

$$c^2 \ll \omega^2 a^2, \qquad (6.B.7)$$

so that, in fact, the tide is inverted—i.e., we have a low tide when the moon is directly overhead. The reason for this is simply the fact that as the moon goes around, it attracts the water toward it, forming a tidal bulge on the earth. This tidal bulge, however, cannot keep up with the moon, and lags behind. Our calculations give a lag of 180°, so that low tide occurs when the moon is directly overhead.

Thus, for an ocean of uniform depth and a moon constrained to orbit exactly over the equator, the equatorial tides would be semi-diurnal and inverted. In fact, we know that the major tides are semi-diurnal, although the presence of variable depth in the ocean and land masses complicates the calculation of real tides considerably. But the main features of the discussion in this section, which involve the effect of the lunar disturbing potential on the long waves in the ocean, will carry through in the more complicated calculations done in later sections.

It should be pointed out that although we have always referred to "lunar forces," in point of fact every body capable of exerting a gravitational attraction at the earth's surface is capable of causing a tide, and, in fact, solar tides are easily seen. This is treated in more detail in Problem 6.1.

C. THE EQUATIONS OF MOTION WITH ROTATION

In the treatment of the equatorial tides in the previous section, two important aspects of the problem of tides have been ignored. One of these, the complicated dependence of the angle Θ on the coordinates of the problem, will be treated in the next section. The other important effects which we must consider are the dynamical consequences of the rotation of the earth. In Section 2.A, we saw that if we went to a coordinate system which was rotating with a body, an extra force appeared. In the static case, this was the familiar centrifugal force. Since the measurement of tides involves moving fluids on the surface of the earth, we will have to expand this concept somewhat.

We know that if a force acts in an inertial system in such a way as to produce an acceleration a_0, then that same force acting in a rotating coordinate system will produce an apparent acceleration given by

$$\mathbf{a} = \mathbf{a}_0 - 2\boldsymbol{\omega} \times \mathbf{v} - \boldsymbol{\omega} \times (\boldsymbol{\omega} \times \mathbf{r}) - \frac{d\boldsymbol{\omega}}{dt} \times \mathbf{r}, \qquad (6.C.1)$$

where ω is the frequency of rotation of the coordinate system. If we are sitting in a coordinate system fixed on the surface of the earth, then we can take $d\boldsymbol{\omega}/dt = 0$. The two "extra" terms in the above equation are then the familiar centrifugal and Coriolis forces. It is customary to treat these terms, which actually arise because of the acceleration of the coordinate system, as forces (usually given some name like apparent or ficticious forces) when we write Newton's laws. Once these extra forces are included, we can easily see that the Euler equation, which is just Newton's second law, becomes

$$\frac{D\mathbf{v}}{Dt} + 2\boldsymbol{\omega} \times \mathbf{v} + \boldsymbol{\omega} \times (\boldsymbol{\omega} \times \mathbf{r}) = -\nabla\left(\frac{P}{\rho} + \Omega\right). \qquad (6.C.2)$$

To understand this equation, consider the system shown in Fig. 6.3. The point P represents the spot at which the tides are being measured, the radius of the earth is taken to be a, and the angles θ and ϕ give the location of P. The length $\tilde{\omega}$ is the perpendicular distance from the axis of rotation (taken to be the z-axis) to P. This somewhat cumbersome notation is, unfortunately, standard for this type of system.

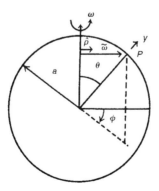

Fig. 6.3. Polar coordinates for the discussion of tidal waves.

In Section 2.A, we treated the centrifugal force by direct integration of the Euler equation for the static case. In this problem, the fluid is in motion relative to the surface of the earth, so we cannot integrate so easily. We can, however, perform an equivalent operation by noting that

$$\boldsymbol{\omega} \times (\boldsymbol{\omega} \times \mathbf{r}) = - \omega^2 \tilde{\omega} \hat{\rho}, \qquad (6.C.3)$$

where $\hat{\rho}$ is a unit vector perpendicular to the z-axis in the direction of P. A simple manipulation (see Problem 6.3) then gives

$$\boldsymbol{\omega} \times (\boldsymbol{\omega} \times \mathbf{r}) = - \nabla\left(\frac{1}{2} \tilde{\omega}^2 \omega^2\right). \qquad (6.C.4)$$

Thus, the centrifugal force term can be written as a gradient, and combined with other terms on the right-hand side of Eq. (6.A.2). If we proceed as in Section 5.A and drop the $(v \cdot \nabla)v$ term in the convective derivative, the Euler equation can be written

$$\frac{\partial \mathbf{v}}{\partial t} + 2\boldsymbol{\omega} \times v = - \nabla\left(\frac{P}{\rho} + \Omega - \frac{1}{2} \omega^2 \tilde{\omega}^2\right), \qquad (6.C.5)$$

where the potential is actually the sum of two terms

$$\Omega = \Omega_e + \Omega_D. \qquad (6.C.6)$$

We have written Ω_e for the potential due to the earth's gravitation, and Ω_D is the disturbing potential due to the presence of the moon derived in Section 6.A.

In the case of long waves (see Section 5.A), we found it very convenient to discuss the y-component on the Euler equation first. The general scheme of things is to solve the y-equation for the quantity whose gradient appears on the right-hand side of Eq. (6.C.5), and then insert this into the remaining equations. The y-equation is

$$\frac{\partial v_y}{\partial t} - 2\omega v_\phi \sin \theta = - \frac{\partial}{\partial y}\left[\frac{P}{\rho} + \Omega_e + \Omega_D - \frac{1}{2} \omega^2 \tilde{\omega}^2\right]. \qquad (6.C.7)$$

If we now invoke the long-wave approximations that were introduced in Section 5.A, we will set the left-hand side of this equation equal to zero. This corresponds to assuming that the motion in the y-direction is slow enough to be regarded as quasi-static. If we then integrate the right-hand side from some arbitrary point y to the point $y = h + \eta$ (which we again take to be the surface of the fluid), we have

$$\frac{P}{\rho} + \Omega_e + \Omega_D - \frac{1}{2} \omega^2 \tilde{\omega}^2 \bigg]_y^{h+\eta} = 0. \qquad (6.C.8)$$

A number of points can be made about this result. First, just as we dropped F_y in Eq. (5.A.11), we will ignore Ω_D with respect to Ω_e in this equation. Second, the quantity $\Omega_e - \frac{1}{2}\omega^2\tilde{\omega}^2$ is the potential which would be felt by a stationary body at the surface of the earth, and is usually referred to as the "apparent gravity." If we expand this quantity at $y = h + \eta$ in a Taylor series about $y = h$, we have

$$\Omega_e - \frac{1}{2}\omega^2\tilde{\omega}^2\Big]_{y=h+\eta} = \Omega_e - \frac{1}{2}\omega^2\tilde{\omega}^2\Big]_{y=h} + \frac{\partial}{\partial y}\left(\Omega_e - \frac{1}{2}\omega^2\tilde{\omega}^2\right)\Big]_{y=h}\eta + \cdots$$
$$= \text{const.} + g\eta. \tag{6.C.9}$$

The final result for the integrated y-component of the Euler equation (Eq. (16.C.8)) is then just

$$\frac{P}{\rho}(y) + \Omega_e(y) - \frac{1}{2}\omega^2\tilde{\omega}^2 = g\eta. \tag{6.C.10}$$

Substituting this result into the right-hand side of Eq. (6.C.5), we find the θ component of the Euler equation to be

$$\frac{\partial v_\theta}{\partial t} - 2\omega v_\phi \cos\theta = -\frac{g}{a}\frac{\partial}{\partial\theta}\left(\eta + \frac{\Omega_D}{g}\right) \tag{6.C.11}$$

and

$$\frac{\partial v_\phi}{\partial t} + 2\omega v_\theta \cos\phi = \frac{g}{\tilde{\omega}}\frac{\partial}{\partial\phi}\left(\eta + \frac{\Omega_D}{g}\right) \tag{6.C.12}$$

for the ϕ-component, where we have dropped terms in v_y.

The remaining equation of motion which must be written down is continuity. In Section 5.A, we saw that it was simpler to derive the equation from the start for the particular geometry in question. The derivation consisted of calculating the amount of fluid in an infinitesimal slice of volume, and then noting that any fluid which enters or leaves the volume must result in a change of height (and therefore a change of η) of the fluid in the volume.

The same technique can be applied for the geometry appropriate to the surface of the earth, although it is a little more difficult to visualize in this case. We can imagine the infinitesimal volume element, which was a simple two-dimensional slice in Section 5.A to be a body extending upward radially from the surface of the earth, so that its height is measured in terms of the coordinate y. In the unperturbed state, this body would be filled to a height h with fluid. Let the perimeter of the body be delineated by arcs (see Fig. 6.4), one corresponding to an infinitesimal increment in θ, and the other to an infinitesimal increment in ϕ. This

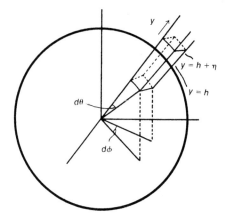

Fig. 6.4. The idea of continuity in polar coordinates.

should be familiar to the reader, since it is the standard volume element in spherical coordinates. With this geometry, it is relatively straightforward to repeat the derivation of Section 5.A to get

$$\frac{\partial \eta}{\partial t} = -\frac{1}{\omega} \left[\frac{\partial}{\partial \theta} (h v_\theta \sin \theta) + \frac{\partial}{\partial \phi} (h v_\phi) \right] \qquad (6.C.13)$$

for the equation of continuity (see Problem 6.4).

D. TIDES AT THE SURFACE OF THE EARTH

In the equatorial theory of the tides, we assumed that both the moon and the point at which the tides were to be observed were on the equator, so that the angle Θ in Fig. 6.2 could be identified with the angle ϕ in our new coordinate system. For the general problem of finding the tides at an arbitrary point on the surface of the earth, this is no longer possible. In fact, if we say that the direction of the radius vector to the moon is given by the angles Δ and α while the radius vector to the point P is given by θ and ϕ (see Fig. 6.5), then

$$\cos \Theta = \cos \Delta \cos \theta + \sin \Delta \sin \phi \cos (\alpha + \phi). \qquad (6.D.1)$$

We can now insert this into the equation for the disturbing potential, Eq. (6.A.4), and put the resulting expression for Ω_D into Eqs. (6.C.11) and (6.C.12) to get the equations governing the tides. Before doing so, however, it will be profitable to discuss the form for the disturbing

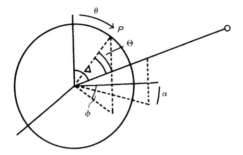

Fig. 6.5.   The angles involved in the general theory of the tides.

potential which results from this manipulation. Recalling that

$$g = \frac{GM_e}{a^2},$$
(6.D.2)

where M_e is the mass of the earth, and defining

$$H = \frac{3}{2} \frac{M_m}{M_e} \frac{a^4}{D^3},$$
(6.D.3)

we have

$$\frac{\Omega_D}{g} = \frac{3}{2} H \left(\cos^2 \Delta - \frac{1}{3} \right) \left(\cos^2 \theta - \frac{1}{3} \right)$$

$$+ \frac{1}{2} H \sin 2\Delta \sin 2\theta \cos (\alpha + \phi)$$

$$+ \frac{1}{2} H \sin^2 \Delta \sin^2 \theta \cos 2(\alpha + \phi)$$

$$= F_1 + F_2 + F_3.$$
(6.D.4)

Thus, we see that the disturbing potential can be thought of as consisting of three separate terms. Since each of these terms plays the same role as the inhomogeneous term in Eq. (6.B.2), it is reasonable to suppose that each is associated with a separate motion of the fluid, and the total motion of the fluid will be the sum of the three separate motions. This property of differential equations is discussed in Appendix E, and we will see some explicit examples later in this section. For the moment, however, let us assume that this is the case and proceed with the discussion.

We see that the time dependences of F_1, F_2, and F_3 are all quite different. If P is fixed, then θ and ϕ do not vary with the time. Since α is the projection of the moon's shadow onto the equatorial plane, we must have

$$\alpha \sim \omega t, \qquad (6.D.5)$$

where ω is the frequency of the earth's rotation. Over the period of a day or so, Δ, the angle of declination of the moon, is approximately constant. Thus, to a first approximation, F_1 is a constant term, F_2 is a term which varies as $\cos \omega t$ and would hence give rise to a once-a-day (diurnal) tide, while F_3 varies as $\cos 2\omega t$, and is associated with the twice-a-day (semi-diurnal) tide.

In fact, we know that the earth's rotation axis is tilted at about 23° with respect to the plane of the moon's orbit, so the angle Δ will have a time dependence whose frequency will be about a month. In addition, the "constant" H contains a factor $1/D^3$, where D is the distance to the moon. D itself changes with time over one lunar revolution, corresponding to the fact that the moon and the earth describe elliptical orbits about their common center of mass. Thus, the simple statements given about time dependences in the above paragraph are not strictly true. It is clear, however, that whatever the time dependence of the angle Δ and the parameter H, they are very slow compared to the time dependence of the angle α. Hence, in calculating the tides due to F_2 and F_3, we can regard both of these as constants, but as constants whose values may change over many periods of the tide, and which must therefore be adjusted before undertaking numerical calculations. This idea is usually expressed by writing

$$F_2 = H'' \cos (\alpha + \phi) \qquad (6.D.6)$$

and

$$F_3 = H''' \cos 2(\alpha + \phi), \qquad (6.D.7)$$

where H'' and H''' are approximately constants. In what follows, we shall ignore the monthly tides associated with F_1, although they are known to exist and have been measured.

We know that in dealing with complicated equations, it is often best to isolate various terms for consideration. We will therefore consider the Euler equation in which Ω_D/g in Eq. (6.D.4) is replaced by either F_2 or F_3, which, for the sake of convenience, we will write as F_i. Let us denote by η_i the displacement of the surface associated with the term F_i. The Euler equations are

$$\frac{\partial v_\theta}{\partial t} - 2\omega v_\phi \cos \theta = -\frac{g}{a} \frac{\partial}{\partial \theta} (\eta_i + F_i) \qquad (6.D.8)$$

and

$$\frac{\partial v_\phi}{\partial t} + 2\omega v_\phi \cos\theta = \frac{g}{\omega}\frac{\partial}{\partial\phi}(\eta_i + F_i). \tag{6.D.9}$$

In general, the t and ϕ dependence of F_i will be of the form

$$F_i \sim e^{i(\sigma_i t + s_i \phi)},$$

where we used the definitions of the sine and cosine in terms of exponentials, and Eq. (6.D.5) which gives the angle α as a function of time. σ_i and s_i are

$$\begin{aligned} \sigma_2 &= \omega, & s_2 &= 1, \\ \sigma_3 &= 2\omega, & s_3 &= 2. \end{aligned} \tag{6.D.10}$$

for F_2 and F_3, respectively.

It becomes natural, therefore, to look for solutions of the form (dropping the subscript i for convenience)

$$\begin{aligned} v_\theta(\theta, \phi, t) &= v_\theta(\theta)e^{i(\sigma t + s\phi)}, \\ v_\phi(\theta, \phi, t) &= v_\phi(\theta)e^{i(\sigma t + s\phi)}, \\ \eta(\theta, \phi, t) &= \eta(\theta)e^{i(Tt + s\phi)}, \end{aligned} \tag{6.D.11}$$

which, upon substitution into Eqs. (6.D.8) and (6.D.9) yields

$$i\sigma v_\theta - 2\omega v_\phi \cos\theta = -\frac{g}{a}\frac{\partial}{\partial\theta}(\eta'_i),$$

$$i\sigma v_\phi + 2\omega v_\theta \cos\theta = is\frac{g}{\omega}\eta'_i, \tag{6.D.12}$$

where we have defined

$$\eta'_i = \eta_i(\theta) + \frac{F_i}{e^{i(\sigma_i t + s_i \phi)}}. \tag{6.D.13}$$

These equations are now algebraic, and therefore quite easy to solve. The net effect of the assumption which we made about the form of the solution, we see, was to reduce the complexity of the Euler equations. Simple algebra (see Problem 6.5) then yields

$$v_\theta(\theta) = \frac{i\sigma}{4m(f^2 - \cos^2\theta)}\left[\frac{\partial\eta'}{\partial\theta} + \frac{s}{f}\eta'\cot\theta\right] \tag{6.D.14}$$

and

$$v_\phi(\theta) = -\frac{\sigma}{4m(f^2 - \cos^2\theta)}\left[\frac{\cos\theta}{f}\frac{\partial\eta'}{\partial\theta} + s\eta'\cos\theta\right], \tag{6.D.15}$$

where we have defined

$$f = \frac{\sigma}{2\omega} \tag{6.D.16}$$

and

$$m = \frac{\omega^2 a}{g}. \tag{6.D.17}$$

(Note that m is not a mass.)

But of course, we must do more than just solve the Euler equations if we are to have a solution. We must solve and satisfy the equation of continuity as well. If we put our assumed forms of the solution into Eq. (6.C.13), we find

$$i\sigma\eta(\theta) = \frac{-1}{a \sin \theta} \left[\frac{\partial}{\partial \theta} (h(\theta)v_\theta(\theta) \sin \theta) + ish(\theta)v_\phi(\theta) \right], \tag{6.D.18}$$

where we have assumed that h, the depth of the ocean at the point P, is a function of θ only. This approximation is not valid for the real earth, of course, any more than the approximation of uniform depth in Eq. (6.B.2) was.

The general problem of the solution of the tides can now be seen to involve solving Eq. (6.D.18), or its more general form which includes a ϕ dependence in the depth, together with the Euler equations, (6.D.14) and (6.D.15). For an arbitrary depth law (by which we mean the dependence of h on θ and ϕ), it is not possible to do this explicitly although it can be done numerically.

There is, however, one depth law which does allow explicit solutions for both η_2 and η_3. Suppose we consider an ocean whose depth is given by

$$h(\theta) = h_0 \sin^2 \theta. \tag{6.D.19}$$

This is actually not a bad approximation to the oceans on the earth—at least it keeps the idea of the oceans at the poles being shallower than those at the equator.

Let us begin by calculating η_2, the diurnal tidal displacement. For this case, we have

$$f = \frac{1}{2}.$$

Since

$$F_2 = H'' \sin \theta \cos \theta \, e^{i(\omega t + \phi)}, \tag{6.D.20}$$

it is natural to assume a form of solution

$$\eta_2' = C \cos \theta \sin \theta \, e^{i(\omega t + \phi)}. \tag{6.D.21}$$

Putting this into the Euler equations in the form of Eqs. (6.D.14) and (6.D.15) quickly yields

$$v_\theta = -\frac{i\sigma C}{m} \qquad (6.D.22)$$

and

$$v_\phi = \frac{\sigma C}{m} \cos \theta.$$

Inserting this into Eq. (6.D.18), recalling the definition in Eq. (6.D.13), we find

$$\eta_2 = \frac{h_0/ma}{1 - 2h_0/ma} H'' \sin \theta \cos \theta\, e^{i(\omega t + \phi)}. \qquad (6.D.23)$$

Turning now to the semi-diurnal tide associated with F_3, we can proceed in analogy to Eq. (6.D.21) to assume that

$$\eta_3' = B \sin^2 \theta\, e^{i(2\omega t + 2\phi)}. \qquad (6.D.24)$$

If we again turn to the Euler equations, we find since $f = \frac{1}{4}$ that

$$v_\theta = \frac{i\sigma B}{m} \cot \theta \qquad (6.D.25)$$

and

$$v_\phi = -\frac{\sigma B}{2m} \left(\frac{1 + \cos^2 \theta}{\sin \theta} \right).$$

Proceeding as before and inserting these into the equation of continuity, we find

$$\eta_3 = -\frac{2h_0}{ma} \frac{H'''}{1 - 2h_0/ma} \sin^2 \theta\, e^{i(2\omega t + 2\phi)} \qquad (6.D.26)$$

for the displacement due to the semi-diurnal tide.

The total displacement at the point P will, of course, be given by

$$\eta_P = \eta_2 + \eta_3, \qquad (6.D.27)$$

so that an observer will see both a daily and a twice daily tide. Comparing Eqs. (6.D.23) and (6.D.26) we would expect these tides to be roughly of equal importance, but this question is examined in more detail in Problem 6.8.

We note that the semi-diurnal tide is still inverted, so that η_3 is negative when the moon is directly overhead. This is not a general result for all $h(\theta, \phi)$, however. For example, calculations of semi-diurnal tides in an ocean of uniform depth give noninverted tides for some latitudes. In Problem 6.7, this problem is dealth with further.

Finally, we know for real tides on the real earth, the diurnal and semi-diurnal tides are not of equal importance. The major tides come twice a day. (The author, born and raised in the Midwest, learned this fact when he began studying tides by listening to late-night radio reports from Norfolk.) Can we understand this feature of the tides on the basis of our simple theories?

In Problem 6.6, we show that the analysis presented above applied to an ocean whose depth is given by

$$h(\theta) = h_0(1 - q \cos^2 \theta) \qquad (6.D.28)$$

yields a diurnal tide for which

$$\eta_2 = \frac{2q(h_0/ma)}{1 - 2q(h_0/ma)} H'' \sin \theta \cos \theta \, e^{i(\omega t + \phi)}. \qquad (6.D.29)$$

We note immediately that for an ocean of uniform depth, where $q = 0$, there is *no* diurnal tide at all. Thus, the diurnal tide exists only insofar as the ocean departs from complete uniformity. Since the oceans are approximately uniform, we would expect that the importance of the diurnal tide should be greatly diminished. This explanation was one of the great triumphs of the Laplace theory. It also explains why no diurnal tides appeared in Section 6.B, when we considered equatorial tides in an ocean of constant depth.

SUMMARY

The net gravitational attraction at the surface of the moon is given by the disturbing potential. This attraction is the cause of the tides. Some simple geometry shows that tides at an arbitrary point will be of three types—a monthly tide, a daily tide, and a semi-diurnal tide. For some simple forms of the depth law for the oceans, it is possible to solve for these tides explicitly, taking into account the rotation of the earth. We find that the semi-diurnal tides are the most important.

PROBLEMS

6.1. For the case of equitorial tides, compare the maximum tide due to the moon with tides due to (a) the sun, (b) Jupiter, and (c) Alpha Centauri.

6.2. Would equatorial tides be inverted on Venus or on Mars (assuming that they had oceans of the same depths as our own)?

6.3. Verify Eq. (6.C.4) and show how it is related to Eq. (2.A.4).

6.4. Consider the problem of continuity in spherical coordinates, as shown in Fig. 6.4.

(a) Show that the amount of fluid in the body at any time is

$$M = \rho a \, d\theta \bar{\omega} \, d\phi \, (h + \eta).$$

(b) Show that the net flux through the walls of length $a \, d\theta$ is

$$\rho \frac{\partial}{\partial \theta} (v_\theta h \bar{\omega}) \, d\theta \, d\phi.$$

(c) Show that the net flux through the walls of length $a \sin \theta \, d\phi$ is

$$\rho \frac{\partial}{\partial \phi} (v_\phi h a) \, d\theta \, d\phi.$$

(d) Hence verify Eq. (6.C.13).

6.5. Verify Eqs. (6.D.14) and (6.D.15).

6.6. For a depth law of the form

$$h(\theta) = h_0(1 - q^2 \cos^2 \theta),$$

show that the diurnal displacement is given by

$$\eta_2 = \frac{2q(h_0/ma)}{1 - 2q(h_0/ma)} H'' \sin \theta \cos \theta \, e^{i(\omega t + d)}.$$

6.7. For the depth law in Eq. (6.D.19), find the smallest value of h_0 such that the tide is not inverted.

6.8. For the depth law of Eq. (6.D.19), calculate the ratio of the maximum values of the diurnal and semi-diurnal tides as a function of longitude. Make a rough sketch of the results.

6.9. Consider the earth to be a sphere of radius a which is covered by an ocean of uniform depth h which is much less than a. Let η be the deviation of the depth of the ocean from uniformity.

(a) Using the methods of Problem 6.4, show that the equation of continuity is

$$\frac{h}{a} \frac{\partial v_\theta}{\partial \theta} + \frac{h}{a} \tan \theta v_\theta + \frac{h}{a \sin \theta} \frac{\partial v_\phi}{\partial \phi} = -\frac{\partial \eta}{\partial t}.$$

(b) If we neglect Coriolis and centrifugal forces, show that the θ- and ϕ-components of the Euler equation are

$$\frac{\partial v_\theta}{\partial t} = -\frac{g}{a} \frac{\partial \eta}{\partial \theta}$$

and

$$\frac{\partial v_\theta}{\partial t} = -\frac{g}{a \sin \theta} \frac{\partial \eta}{\partial \phi},$$

respectively.

(c) Hence show that

$$\frac{1}{a^2 \sin \theta} \frac{\partial}{\partial \theta} \left(\sin \theta \frac{\partial \eta}{\partial \theta} \right) + \frac{1}{a^2 \sin^2 \theta} \frac{\partial^2 \eta}{\partial \phi^2} = \frac{1}{g^2 h^2} \frac{\partial^2 \eta}{\partial t^2}.$$

6.10. (a) For the wave equation derived in Problem 6.9, show that using the technique of separation of variables outlined in Appendix F, the solution for η will be of the form

$$\eta \propto Y_{lm}(\theta, \phi) e^{i\omega t},$$

where Y_{lm} is the spherical harmonic defined in Appendix F.

(b) If we define $\beta = (a/\omega c)^2$, show that the only solutions which are possible are those for which

$$\beta = l(l + 1).$$

(*Hint*: Consider the case where $\cos \theta = +1$, and use the recursion relation for Legendre polynomials given in Appendix F to show that η will be infinite unless the Legendre series terminates.)

(c) These allowed frequencies are associated with the *normal modes of oscillation*. Calculate the frequencies for the first four modes for the earth.

(d) Consider a plane through the earth at $\phi = 0$. Sketch the value of η as a function of θ for the first few normal modes.

6.11. (a) Continuing with the example of the flooded earth in the previous problems, show that if a disturbing potential is present, the Euler equations in Problem 6.9 will have a term

$$\frac{1}{r} \frac{\partial}{\partial \theta} (\Omega_D) \quad \text{and} \quad \frac{1}{r \sin \theta} \frac{\partial}{\partial \phi} (\Omega_D)$$

added, respectively, to the θ and ϕ equations.

(b) Derive the new wave equation corresponding to the new Euler equations, and show that it can be written in the form

$$\frac{1}{a^2 \sin \theta} \frac{\partial}{\partial \theta} \left(\sin \theta \frac{\partial \eta}{\partial \theta} \right) + \frac{1}{a^2 \sin^2 \theta} \frac{\partial^2 \eta}{\partial \phi^2} - \frac{1}{g^2 h^2} \frac{\partial^2 \eta}{\partial t^2} = \sum_{l,m} a_{lm} Y_{lm}(\theta, \phi) e^{i\omega_l t}.$$

(*Hint*: Use the expansion in spherical harmonics discussed in Appendix F.) Evaluate a_{lm} for the first three values of l.

(c) Using the theory of inhomogeneous equation outlined in Appendix E, find the particular solution to this equation for the first three l values. Sketch the results.

REFERENCES

H. Lamb, *Hydrodynamics* (cited in Chapter 1) has an excellent discussion of the theory of the tides. The following texts are also quite useful.

R. A. Becker, *Introduction to Theoretical Mechanics*, McGraw-Hill, New York, 1954.

In Chapter 11, there is a readable and complete discussion of equations of motion in accelerated frames, and of the Coriolis and centrifugal forces.

Walter Kauzmann, *Quantum Chemistry*, Academic Press, New York, 1957.

Chapter 3 of this text contains a very nice description of the use of spherical harmonics applied to the problem of tides on the earth.

William S. von Arx, *An Introduction to Physical Oceanography*, Addison-Wesley, New York, 1962.

An excellent descriptive text on the motion of the oceans, currents, and waves, along with a discussion of how measurements are made.

C. Eckart, *Hydrodynamics of Oceans and Atmospheres*, Pergamon Press, New York, 1962.

Contains an excellent discussion of the tidal equations with rotation, and of the actual structure of the ocean.

7

Oscillations of Fluid Spheres: Vibrations of the Earth and Nuclear Fission

> He felt the earth move out and away from under them.
>
> ERNEST HEMINGWAY
> *For Whom the Bell Tolls*

A. FREE VIBRATIONS OF THE EARTH

In Section 2.D, we say that for some purposes it is reasonable to treat the earth as if it were a uniform liquid. If we ignore the rotation of the earth for the discussion in this section, then the equilibrium configuration of the earth would be a sphere. It is reasonable to ask what would happen to such a sphere if, for some reason, it were slightly deformed (e.g., by an earthquake) and then allowed to respond. We will show in this section that a liquid sphere would be expected to perform oscillations about its equilibrium configuration. This phenomenon, similar to the ringing of a bell, has recently been measured by geophysicists.

If the earth in its unperturbed state is a sphere of radius a and density ρ, then in a perturbed state, the distance from the center to the perturbed surface will be (see Fig. 7.1)

$$r = a + \zeta(\theta, \phi). \tag{7.A.1}$$

It is always possible to expand the function $\zeta(\theta, \phi)$ in terms of spherical harmonics (see Appendix F)

$$\zeta(\theta, \phi) = \sum_l \sum_m a_{lm} Y_{lm}(\theta, \phi) = \sum_l \zeta_l, \tag{7.A.2}$$

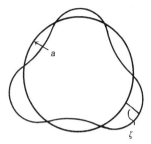

Fig. 7.1.   The initial distortion of a liquid sphere.

where we have incorporated the sum over m into the definition of ζ_l.

As in Section 4.B, the equation which determines the velocity potential is just

$$\nabla^2 \phi = 0, \tag{7.A.3}$$

which for a spherical geometry has the solution (see Appendix F)

$$\phi = \sum_l \sum_m \left[A_{lm} \left(\frac{r}{a}\right)^l + B_{lm} \left(\frac{a}{r}\right)^{l+1} \right] Y_{lm}(\theta, \phi).$$

Since we want ϕ to be bounded at $r = 0$, we must have $B_{lm} = 0$ (this corresponds to using the "bottom" boundary condition in Eq. (5.B.6)—in this case the condition is that at the origin the velocity is finite). Thus, we can write

$$\phi = \sum_l \sum_m A_{lm} Y_{lm}(\theta, \phi) \left(\frac{r}{a}\right)^l = \sum_l \left(\frac{r}{a}\right)^l s_l, \tag{7.A.4}$$

where s_l is defined in a manner similar to ζ_l.

The boundary condition at the outer surface is given by the Euler equation as in Eq. (5.B.10) to be

$$\frac{\partial \zeta}{\partial t} = \frac{\partial \phi}{\partial r}, \tag{7.A.5}$$

which can be written

$$\frac{\partial \zeta_l}{\partial t} = \frac{l}{a}, \tag{7.A.6}$$

where, as in the development of surface waves in Section 5.B, we have evaluated the boundary equation at the unperturbed surface.

We now turn to the problem of finding whether or not we can find wavelike solutions for ζ, the deviation of the surface of the sphere from equilibrium at a given point. If such solutions are found, then an observer at that point would observe the oscillations in the earth that we are discussing. As in the development of surface waves [see Eq. (5.B.2)], we use the Euler equation in the form

$$\frac{P}{\rho} + \frac{\partial \phi}{\partial t} + \Omega = 0 \qquad (7.A.7)$$

to determine the existence of oscillations and their frequency. In order to use this equation, however, we have to find the potential due to a distorted sphere. We will solve the general problem of finding the potential at a point r just above the unperturbed surface of the sphere, and later let $r \to a$ (as is appropriate for our procedure of evaluating all boundary conditions at the unperturbed surface).

We shall find the problem to be considerably simplified if we break the potential into two parts (see Fig. 7.2)—one the potential at r due to an unperturbed sphere of radius a, and the other the potential at r due to a thin spherical shell of variable density $\rho_s(\theta, \phi)$. Later, we shall see how to relate this variable density to the displacement of the surface, ζ. For the moment, however, we simply note that the density of the shell can be either positive or negative, depending on whether the actual surface is above or below the unperturbed surface at a given point. Thus,

$$\Omega(r) = -\frac{4}{3}\pi a^3 \rho \frac{G}{r} + \Omega_{\text{shell}}, \qquad (7.A.8)$$

where the first term of the right-hand side is the potential at r due to the sphere, and the second term (still to be calculated) represents the

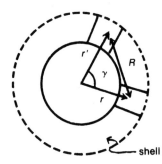

Fig. 7.2. Coordinates for breaking the sphere into a central core plus a shell.

potential at r due to the rest of the shell. This is

$$\Omega_{\text{shell}} = -G \int \frac{\rho_s(\theta', \phi')\, ds}{R(\theta', \phi')} = -Ga^2 \int \frac{\rho_s(\theta', \phi')}{R(\theta', \phi')} d\,(\cos\theta')\, d\phi, \quad (7.A.9)$$

where, because we are considering only small vibrations, and hence very thin shells, we can regard $\rho_s(\theta', \phi')$ as a surface mass density. The surface density, being a function of θ' and ϕ', can be expanded in terms of spherical harmonics, just as we expanded ζ in Eq. (7.A.2).

$$\rho_s(\theta', \phi') = \sum_L \sum_M \rho_{LM}^{(s)} Y_{LM}(\theta', \phi') = \sum_L \rho_L^{(s)}. \quad (7.A.10)$$

It is a standard mathematical result that the term $1/R$ which appears in Eq. (7.A.9) can be written as

$$\frac{1}{R(\theta', \phi')} = \sum_l \frac{(r')^l}{r^{l+1}} P_l(\cos\gamma), \quad (7.A.11)$$

where γ is the angle between r and r' (see Fig. 7.2).

If we take these results and put them back into Eq. (7.A.9), we can use the results of Problem 7.1 to carry out the integrals over the angles θ' and ϕ'. Letting $r = a$, we find that Ω_{shell} evaluated at the unperturbed surface of the sphere is

$$\Omega_{\text{shell}} = -Ga \sum_l \left(\frac{r}{a}\right)^l \frac{4\pi}{2l+1} \rho_L^{(s)}$$
$$= \sum_l \Omega_l. \quad (7.A.12)$$

All we need to do now is determine the surface density of the shell $\rho_L^{(s)}$, and we will have the potential due to a distorted sphere. Consider Fig. 7.3. The shaded area represents the excess mass in the surface element due to the distortion of the surface. The amount of excess mass is just $\rho\zeta\, d\sigma$,

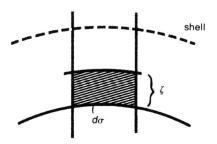

Fig. 7.3. Situation at the surface of the distorted sphere.

which we must equate to $\rho^{(s)} d\sigma$, the mass in a surface element of the shell. Thus,

$$\Omega_l = -\frac{4\pi\rho Ga}{2l+1}\,\zeta_l, \tag{7.A.13}$$

where Ω_l is defined in Eq. (7.A.12). This means that

$$\Omega = -\frac{ga^2}{r} - \sum_l \frac{3g}{2l+1}\,\zeta_l \tag{7.A.14}$$

is the total potential at the point r. Letting r in the first term be $a + \zeta$, and then keeping only first-order terms in ζ, we find

$$\Omega = -ga + g\sum_l \frac{2(l-1)}{2l+1}\,\zeta_l, \tag{7.A.15}$$

where we have used the identity $\frac{4}{3}\pi a\rho = g$. We leave as an exercise for the reader the problem of why we set $r = a$ in the calculation of Ω_{shell}, but had to set $r = a + \zeta$ in Eq. (7.A.14).

We are now ready to use the Euler equation at the surface to determine the equation for ζ_l. Using our prescription of evaluating all terms at the unperturbed surface, and using the condition that the pressure at the surface must be a constant, we find

$$\text{const.} = \sum_l \left[\frac{\partial s_l}{\partial t} - g\frac{2(l-1)}{2l+1}\,\zeta_l\right]. \tag{7.A.16}$$

Differentiating this equation with respect to time, and using the boundary condition in Eq. (7.A.6), we find

$$\frac{\partial^2 \zeta_l}{\partial t^2} + \left[\frac{l}{a}\frac{2(l-1)}{2l+1}g\right]\zeta_l = 0, \tag{7.A.17}$$

which is, indeed, the equation of a harmonic oscillator, with frequency of oscillation given by

$$\omega^2 = \frac{l}{a}\frac{2(l-1)}{2l+1}g. \tag{7.A.18}$$

An observer at the surface, then, will see oscillations corresponding to the above frequencies if for some reason the surface of the earth is ever distorted.

Of course, in a general excitation, we would expect all possible frequencies to be excited, and the actual displacement of the surface would be some sort of series, where the frequencies of each term in the series are given by the above equation. Let us look at the first few terms in

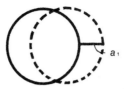

Fig. 7.4. The distortion corresponding to $l = 1$.

such a series (corresponding to the lowest values of l). For $l = 0$, $\omega = 0$ and no time-dependent displacements would be observed.

For $l = 1$, $\omega = 0$ also. This type of displacement of the surface would correspond to $r = a + a_1 \cos \theta$ which would correspond to an overall displacement of the sphere, and could not be detected by an observer at the surface (see Fig. 7.4).

Thus, the lowest observable oscillation would correspond to $l = 2$, or $r = a + a_2(3 \cos^2 \theta - 1)$ (we will ignore the dependence on ϕ for simplicity). This corresponds to a distortion such as that shown in Fig. 7.5, which has a frequency

$$\omega_2{}^2 = \frac{16}{15}\,\pi G \rho.$$

This corresponds to a time between pulses at the surface of the earth of about

$$\tau_2 = \frac{2\pi}{\omega_2} \approx 45 \text{ min,}$$

which is close to the 3–60 minute pulses observed after the Chilean earthquake of 1960!

B. THE LIQUID DROP MODEL OF THE NUCLEUS

Throughout this text, we have emphasized the fact that hydrodynamics is a subject which can be applied over a wide range of physical phenomena. Perhaps nowhere is that fact so surprising as in the realization that some of the earliest ideas about the atomic nucleus were

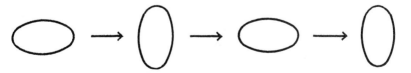

Fig. 7.5. The lowest observable oscillation for a liquid sphere.

based on concepts of fluid mechanics. It is to these models that we now turn our attention.

It may seem strange that the classical theory of fluids should have anything to do with nuclear effects, but actually it is not. The problem of describing a nucleus made up of many interacting nucleons is in many ways similar to the problem of describing a gas made up of many interacting particles. If one does not want to get involved in the impossible problem of describing the motion of each particle in detail, one treats the system as an ensemble, and discusses only the gross properties, ignoring the detailed structure as much as possible. In the case of the liquid, one uses thermodynamics or fluid mechanics. Since a fluid is the simplest system in which this averaging process is done, it is natural to try to approximate any system with a complex internal structure by a fluid. The liquid drop model represents such a zero-order approximation to the behavior of large nuclei.

In the discussion of surface tension, we showed how the existence of an attractive force between the constituent particles of a liquid gave rise to a surface force. A nucleus is made up of protons and neutrons, so that if there were no forces present other than electromagnetic ones, the nucleus would have to fly apart because of the Coulomb repulsion between protons. The existence of nuclei is thus evidence for the existence of short-range attractive forces between the nucleons. (These are the "strong interactions" which constitute one of the major fields of investigation in modern physics.) Such a force would, of course, give rise to a surface tension in the nuclear "fluid." The stability of the nucleus is thus seen to be a result of the competition between the Coulomb electrostatic forces, which tend to blow the nucleus apart, and the strong interactions giving rise to a surface tension, which tends to hold the nucleus together. (These two forces play similar roles to gravity and centrifugal force, whose competition was the main point of investigation in our study of stars in Chapter 2.)

In our discussion of stability in Chapter 3, we saw that one way to decide whether a system is stable against some perturbation is to see whether that perturbation increases or decreases the energy of the system. Therefore, let us consider the stability of nuclei by considering the deformation of a nucleus whose radius when undisturbed is a, and whose strong interactions give rise to a surface tension T. Let us take an arbitrary deformation of the surface such as that shown in Fig. 7.1, so that the distance from the center to the surface is just

$$R = a\left(1 + \sum_l \sum_m a_{lm} Y_{lm}(\theta, \phi)\right). \tag{7.B.1}$$

This distortion has two competing effects. First, by increasing the surface area, we increase the surface energy, which is given by

$$E_S = TS, \qquad (7.B.2)$$

where S is the total surface area, and, second, we move the charges farther apart from each other, so we decrease the Coulomb effects. The interplay between these two effects will determine the stability of the system.

In Problem 7.3, it is shown that the surface area of a sphere deformed according to Eq. (7.B.1) is

$$S = 4\pi a^2 \left(1 + \frac{1}{8\pi} \sum_l |a_{lm}|^2 (l-1)(l+2)\right). \qquad (7.B.3)$$

If we write the surface energy of the undeformed sphere as

$$E_S^0 = 4\pi a^2 T, \qquad (7.B.4)$$

then the change in surface energy accompanying deformation is just

$$\Delta E_S = \frac{E_S^0}{8\pi} \sum_l |a_{lm}|^2 (l-1)(l+2). \qquad (7.B.5)$$

For the purpose of calculating the Coulomb energy of the deformed sphere, we assume that the total charge of the nucleus (which we shall call Ze, where e is the charge on a single proton) is spread out uniformly over the sphere, so that the charge density is just

$$\rho = \frac{3Ze}{4\pi a^3}. \qquad (7.B.6)$$

As in the problem in the preceding section in which the potential of a deformed sphere was calculated, we shall replace the deformed sphere by a sphere of radius a and a spherical shell whose thickness is small compared to the radius of the sphere (see Fig. 7.2). This problem is similar in many respects to the calculation of the potential of the deformed earth in Section 7.A, but in calculating stability in the way we are doing it, we will be concerned with the energy of a charge distribution in the potential, and not in the potential itself.

The Coulomb energy can be written

$$E_C = \int \rho \Omega \, dV, \qquad (7.B.7)$$

where the integration is understood to extend over the entire deformed sphere. We will find it easier to treat the system as if the thin shell and the sphere were two separate entities. In this case, the Coulomb energy would be made up of three terms: the self-energy of the sphere, given by

$$E_1 = \frac{1}{2} \int_{\text{sphere}} \rho \Omega \, dV = \frac{3}{5} \frac{Z^2 e^2}{a^2}; \qquad (7.B.8)$$

the self-energy of the shell, given by

$$E_2 = \frac{1}{2} \int_{\text{shell}} \rho_S \Omega_S \, dV; \qquad (7.B.9)$$

and the interaction energy between the sphere and the shell, given by

$$E_3 = \int_{\text{shell}} \rho_S \Omega \, dV, \qquad (7.B.10)$$

where we have written the potential of the sphere as Ω, the potential of the shell as Ω_S, and the density of the shell (see Fig. 7.3) as ρ_S.

The difference between the Coulomb energy in the undistorted state and the distorted state is then

$$\Delta E_C = E_2 + E_3. \qquad (7.B.11)$$

The calculation of E_3 is relatively simple. The potential at the shell due to the sphere is just

$$\Omega = \frac{Ze}{r}, \qquad (7.B.12)$$

so that

$$E_3 = -\frac{3Z^2 e^2}{4\pi a^3} \int d\omega \int_\omega^R r' \, dr', \qquad (7.B.13)$$

where $d\omega$ represents the integral over the solid angle. Note that in the event $R < a$, the integral over r' will change sign, so that we need not worry about whether the perturbation pushes R out or pulls it in. Carrying out the integral over the radial variable,

$$\begin{aligned}
E_3 &= -\frac{3Z^2 e^2}{4\pi a^3} \int d\omega \cdot \frac{1}{2} a^2 \left(1 + 2 \sum_{l,m} a_{lm} Y_{lm}(\theta', \phi') \right. \\
&\quad + \left. \left(\sum_{l,m} a_{lm} Y_{lm}(\theta', \phi') \right)^2 - 1 \right) \\
&= -\frac{3Z^2 e^2}{8\pi a} \sum_{l,m} |a_{lm}|^2, \qquad (7.B.14)
\end{aligned}$$

where we have used the orthogonality properties of the spherical harmonics to eliminate the linear term in a_{lm} and to collapse the double sum in the quadratic term (see Appendix F).

The calculation of E_2, the self-energy of the shell, can be split into two parts—the calculation of Ω_S, the potential at a point in the shell due to the rest of the shell, and then the calculation of E_2 itself. From Fig. 7.6, we see that we can write

$$\Omega_S = \rho \int d\omega' \int_a^{R(\theta',\phi')} \frac{r'^2 \, dr'}{|r - r'|}. \tag{7.B.15}$$

We see that Ω_S will depend on $R\text{-}a$, hence will be linear in the small parameter a_{lm}. Since in the calculation of E_3, Ω_S will appear inside another integral which will depend on $R\text{-}a$, it will be sufficient to keep only lowest-order terms in the above expression.

Proceeding as in the steps leading to Eq. (7.A.12), we find

$$\Omega_S(r) = \frac{3Ze}{a} \sum_{l,m} \frac{a_{l,m}^*}{2l+1} Y_{lm}^*(\theta, \phi), \tag{7.B.16}$$

so that

$$E_2 = \frac{9}{8} \frac{(Ze)^2}{\pi a''} \sum_{l,m} \int d\omega \frac{a_{lm}^*}{2l+1} Y_{lm}^*(\theta, \phi) \int_a^{R(\theta,\phi)} r^2 \, dr$$

$$= \frac{9}{8} \frac{Z^2 e^2}{\pi a} \sum_{l,m} \frac{|a_{lm}|^2}{2l+1}. \tag{7.B.17}$$

Combining Eqs. (7.B.3), (7.B.14), and (7.B.17), we find that the total energy change in the system when an infinitesimal deformation takes place is

$$\Delta E = \sum_{l,m} |a_{lm}|^2 \left[Ta^2 \frac{(l-1)(l+2)}{2} - \frac{3Z^2 e^2}{4\pi a^3 T(2l+1)} \right]. \tag{7.B.18}$$

This will be positive or negative, depending on whether the second term (representing the Coulomb energy) is greater or less than the first

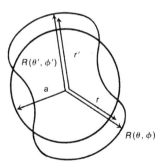

Fig. 7.6. Coordinates for calculating a Coulomb potential for deformed nuclei.

(representing the surface tension). This is what we expected when we remarked earlier that the stability of the system would depend on the interplay between these two forces.

Let us examine this stability criterion as a function of the total number of protons and neutrons in the nucleus. We can write

$$A = \frac{4}{3} \pi a^3 \rho_A,$$

where ρ_A is the density of nucleons in the nucleus. This leads us to expect that the radius of the nucleus, a, should be related to the nuclear number A,

$$a = r_0 A^{1/3}. \tag{7.B.19}$$

This is, in fact, the observed law of nuclear size, and the constant r_0 is generally given a value of about 1.2×10^{-13} cm. For $l = 2$ deformations, this means that the expression for ΔE in Eq. (7.B.18) will be positive, and hence the nucleus will be stable, only if

$$\frac{Z^2}{A} < \frac{40 \pi r_0^3}{3e^2} T. \tag{7.B.20}$$

We know, of course, from the discussion of Chapter 3 that if the system is unstable against one perturbation, then it will not be able to survive in nature.

Thus, the nucleus will be unstable if the relative amount of protons, Z, becomes large compared to the number of protons and neutrons, A, which give rise to the surface tension. In Problem 7.4, it is shown that this leads to a prediction for the largest stable nucleus which is possible in nature. If the stability criterion is not met, then we expect that the nucleus will undergo large oscillations and eventually break up. This is known as *spontaneous fission*, and will be discussed in the next section. It is one process which gives rise to natural radioactivity.

Of course, spontaneous fission is only one kind of instability that a nucleus can have, and only a few nuclei in nature actually exhibit it. Other kinds of instabilities which would break up a nucleus are processes in which the nucleus would emit any of a number of particles. Such processes must be treated quantum mechanically, however, and are not included in the liquid drop model.

Finally, we note that the stability criterion in Eq. (7.B.20) can be written

$$x = \frac{E_1}{2E_s^0} < 1,$$

where E_1 and E_s^0 are given in Eqs. (7.B.8) and (7.B.4). The parameter x is called the *fissionability parameter*, and is sometimes used in discussions of fission.

C. NUCLEAR FISSION

The problem of the fission of heavy elements has, until very recently, defied theoretical analysis. Yet the use of the fission process in reactors has been widespread. Let us use the liquid drop model of the nucleus developed in the previous section to see if we can come to some qualitative understanding of how energy can be derived from fission.

Let us consider what happens when, for some reason, a nucleus is split up. Remember that *any* nucleus can be split up. The stability criterion just tells us which nuclei will not break up spontaneously. We can ask first what kinds of breakup are energetically favored; i.e. which possible final state has the lowest energy. Let us assume for simplicity that the nucleus breaks up into two fragments, one with N nucleons, and the other with A-N. Let us assume that the final product is the two spheres separated by a great distance. Let us also, as a first approximation, ignore the Coulomb energy in the final state [this will be a good approximation unless we are close to the stability limit (see Problem 7.6)]. Then the final energy of the system after the split up will be

$$E_f = 4\pi r_0^2 T[N^{2/3} + (A - N)^{2/3}], \qquad (7.C.1)$$

so that the net energy change is

$$\Delta E = 4\pi r_0^2 T A^{2/3} - E_f.$$

This will be a minimum when

$$\frac{\partial(\Delta E)}{\partial N} = 0 = -N^{-1/3} + (A - N)^{-1/3}$$

or

$$N = \frac{A}{2}. \qquad (7.C.2)$$

Thus, the liquid drop model predicts that when a nucleus breaks up, it should split into two equal-sized fragments. This is actually not the case (e.g., when uranium undergoes fission, the end products are clustered so that when one fragment is around $A = 90$ the other is around $A = 140$). This is one of the main difficulties of the liquid drop model—one of its failures. However, the question of why nuclei should go to unequal fragments has been the subject of a long investigation in the theory of heavy nuclei, and the solutions to the problem which have been advanced depend in a very critical way on details of the quantum mechanics of many body systems.

We will still use the model, however, because although it is wrong in some details, it nevertheless reproduces the general features of nuclear structure quite well in a very simple way.

Let us suppose that a nucleus splits up, then, into two equal fragments, each with half the protons and neutrons of the parent nucleus. What is the final energy of the system?

$$E_f = 8\pi r_\theta^2 T \left(\frac{A}{2}\right)^{2/3} - \frac{6}{5} \frac{e^2(Z/2)^2}{r_0(A/2)^{2/3}}$$

$$= 2^{1/3} E_S{}^0 + \frac{E}{2^{1/3}}. \qquad (7.C.3)$$

Thus, the energy associated with the breakup of the system can be written in terms of the fissionability parameter as

$$\Delta E_B = E_S^0(2^{1/3} - 1) + E_1 \left(\frac{1}{2^{1/3}} - 1\right)$$

$$= E_S^0(0.26 - 0.74x), \qquad (7.C.4)$$

so that when $x > 0.35$ the system can go to a final state of lower energy than the original state. But this is confusing, because we have shown above that the system is stable against small perturbations until $x > 1$. How can these two seemingly contradictory results be reconciled?

The answer, of course, is that the results on stability tell us what happens when *small* perturbations are applied to the system. However, in order for breakup to occur, the perturbations must be very large indeed. Schematically, we can imagine the total energy of the system as a function of perturbation parameter to look like Fig. 7.7. For small perturbations, the system is stable. If an amount of energy $E_{fission}$ is added to the system, however, it will be able to overcome the potential barrier, and

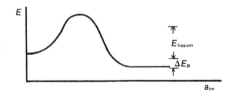

Fig. 7.7.   The energy surface for a fissionable nucleus.

fall to the state discussed above, which has an energy ΔE_B below the initial state. This is called *induced fission.*

Most of the current research on fission has to do with mapping out the complicated energy surfaces which correspond to various deformations of the nucleus, and then trying to decide how fission will proceed for real nuclei.

Induced fission is the basic principle by which a fission reactor works. If the nucleus in question is U_{235}, then a neutron striking the nucleus can supply the energy needed to put the nucleus over the top of the potential barrier. The energy released is then ΔE_B (see Problem 7.5). Since some of this energy is released in the form of neutrons, which can, in turn, initiate further reactions, it is possible to sustain a continuous fission process from which energy can be extracted.

This discussion illustrates an important point about hydrodynamics (and, indeed, about any system described by nonlinear equilibrium). The behavior of the system close to equilibrium need not be related simply to the behavior of the system far from equilibrium. This aspect of the physical world is only now beginning to be explored, and very little is known about it at present.

SUMMARY

Application of the principles of fluids which were developed in previous chapters to spherical fluid systems leads to two interesting predictions. First, a fluid (such as the earth) acting under the influence of its own gravity will execute periodic vibrations about equilibrium if deformed and then released. Second, a charged fluid under the influence of surface tension (such as a nucleus) will fission spontaneously for certain values of the charge. This was used to discuss the process of nuclear fission, which is an example of a process in which deviations from the small perturbation, linear theory which we have been presenting are important.

PROBLEMS

7.1. Look up the addition theorem for spherical harmonics, and use it, together with the properties of the spherical harmonics discussed in Appendix F, to show that

$$\int Y_{LM}(\theta', \phi') P_l(\cos \gamma) d(\cos \theta) \, d\phi = \delta_{l,L} \frac{4\pi}{2L+1} Y_{LM}(\theta, \phi).$$

7.2. Given that the gravitational attraction at the surface of the moon is approximately $\frac{1}{6}$ that at the earth, estimate the period of the $l = 2$ free oscillations of the moon. How do they compare with those of the earth?

7.3. Let us consider the surface of a deformed sphere.

(a) From Eq. (7.B.1), show that

$$dR - \frac{1}{R}\frac{\partial R}{\partial \theta} R \, d\theta - \frac{1}{R \sin \theta}\frac{\partial R}{\partial \phi} R \sin \theta \, d\phi = 0.$$

Hence show that the direction cosines of the deformed surface are

$$\frac{1}{E}, \qquad \frac{-\dfrac{1}{R}\dfrac{\partial R}{\partial \theta}}{E}, \qquad \frac{-\dfrac{1}{R \sin \theta}\dfrac{\partial R}{\partial \phi}}{E},$$

where

$$E^2 = 1 + \frac{1}{R^2}\left(\frac{\partial R}{\partial \theta}\right)^2 + \frac{1}{R^2 \sin^2 \theta}\left(\frac{\partial R}{\partial \theta}\right)^2.$$

(b) Hence show that the change in surface area of an infinitesimal volume element is

$$\frac{R^2}{E}\left(1 - \frac{1}{R}\frac{\partial R}{\partial \theta} - \frac{1}{R \sin \theta}\frac{\partial R}{\partial \phi}\right).$$

(c) Integrate to obtain the surface area in the form

$$4\pi a^2 \left(1 + \frac{1}{8\pi}\sum_{l,m}(l-1)(l+2)|a_{l,m}|^2\right).$$

7.4. Find a good value to the surface tension T of a nucleus, and calculate the largest value of Z^2/A which a nucleus can have and still be stable. How does this compare to the actual stability of heavy elements?

7.5. Calculate the energy which will be released if the nucleus U_{235} is made to undergo fission, assuming that the liquid drop model is correct in stating that the final state will be two identical nuclei. How does this compare to the actual value of this number?

7.6. Show that including the Coulomb effect in Eq. (7.C.1) will not affect the conclusion of Eq. (7.C.2) for heavy nuclei. (*Hint*: What is the relation between A and Z around uranium?)

7.7. Verify the expression for E_1 in Eq. (7.B.8).

7.8. Suppose that the earth had a total charge Q spread uniformly through its volume. How would Eq. (7.A.17) be altered? Are there values of Q for which the frequencies of vibration will be complex, and therefore represent an instability? Relate this to the results of Section 7.B.

REFERENCES

K. E. Bullen, *An Introduction to the Theory of Seismology*, Cambridge, U.P., 1965.
Chapter 14 gives the theory of oscillation for a solid earth, and a survey of observations.
Lawrence Willets, *Theories of Nuclear Fission*, Clarendon Press, Oxford, 1964
A survey of nuclear fission. This should give an overview of the field.
M. Brack, J. Damgaard, A. S. Jensen, H. C. Pauli, V. M. Strutinsky, C. Y. Wong, *Reviews of Modern Physics* **44**, 320 (1972).
A review of the latest ideas in the theory of fission.
I. Prigogine, G. Nicolis, and A. Babloyantz, *Physics Today* **25**, numbers 11 and 12 (1972).
They give a discussion of how a living system which is far from the equilibrium of its consituents might arise by processes similar to that considered in Section 7.C.

8

Viscosity in Fluids

Slow as molasses in January

Southern folk saying

A. THE IDEA OF VISCOSITY

Up to this point, we have ignored many of the properties of real fluids which might serve to complicate our considerations of simple systems. We have argued that this is a valid way to proceed in many cases. As might be expected, however, there are many phenomena for which the "ideal fluid" will simply not provide an adequate description.

In an ideal fluid, the only way in which a force can be generated or, equivalently, in which momentum can be transferred, is through the pressure gradient. On the atomic level, this corresponds to collisions in which the momentum of a molecule in the direction of the force is reversed. Clearly, a force of this type must always be normal to the surface on which it is being exerted. In addition, if we were somehow able to reach into an ideal fluid and apply a force to a single fluid element, there would be nothing other than pressure gradients to oppose the motion of the element, so that it could be quickly accelerated.

To see the shortcomings of this description of a fluid, consider the following example: Let there be a fluid of depth h which is not moving. Let another layer of identical fluid be flowing across the top of the stationary layer at a velocity v. For a classical ideal fluid, the fluid in the upper layer will keep moving indefinitely, even if no forces are acting on it. Our intuition tells us, however, that in a real situation, the top layer would eventually slow down and stop. This means that there must be some way of exerting forces which are different from the pressure, and which act *along* a surface, rather than normal to it.

The term usually used to describe such a situation is that the fluid is capable of exerting a *shear force*, in addition to the pressure. The phenomenon associated with this force is called *viscosity*.

To understand how viscosity works at the atomic level, consider a collision between two atoms in the above example. If only pressure forces could be exerted, then momentum transfers could occur only in a direction normal to the interface between the fluids, and the momentum of each atom along the interface would have to remain constant (essentially, the atom in the moving fluid would retain, on the average, a velocity v). When we put things this way, it is clear that the assumptions associated with ideal fluids are rather artificial. Suppose we thought about a more realistic atomic picture, in which momentum could be transferred in any direction. Then the atoms in the lower layer would, on the average, be speeded up by collisions, while the atoms in the upper layer would, on the average, be slowed down. The net result would be that the relative velocity between the two layers would be reduced (eventually) to zero. This mechanism is similar to the phenomenon of friction in mechanics.

In order to come to some basic understanding of viscosity, let us return to the derivations of the Euler equation in Chapter 1, in which Newton's second law of motion was applied to an infinitesimal element of the fluid to give the equation

$$\frac{\partial}{\partial t}(\rho v_i) = -\frac{\partial}{\partial x_k}(P\delta_{ik} + \rho v_i v_k). \tag{8.A.1}$$

From the point of view of the volume element on which the various forces (pressure, gravity, etc.) are acting, the existence of viscosity will be an additional way in which the momentum of the element can be changed, or, by Newton's second law, an additional force. To see why this should be so, consider an element in the moving fluid we discussed earlier. Because of the collisions between moving and stationary atoms, it would experience a net deceleration. To an observer on the element who knew nothing of atomic structure, this would appear to be due to some sort of internal force generated within the fluid, just as the frictional force generated when a block of wood slides across a table slows down the block. (As a matter of historical interest, a common way of thinking about viscosity in classical terms is to imagine the fluid flow as being made up of a series of sheets sliding over each other, and viscosity as being the friction between the sheets.)

The existence of this extra force, or momentum transfer, means that there must be an additional term in the Euler equation. For the sake of definiteness, we will treat viscosity as a force, and put it on the right-hand side of Eq. (8.A.1), but we could just as well treat it as a momentum change, and put in on the left-hand side. Is there anything we can say about the form that this extra term in the Euler equation must take on general grounds? It turns out that there is a great deal that can be said.

The first thing that we note about the Euler equation in the form

$$\frac{\partial}{\partial t}(\rho v_i) = -\frac{\partial}{\partial x_k}\Pi_{ik} + F_i \qquad (8.A.2)$$

is that we can always write the extra force as

$$F_i = \frac{\partial}{\partial x_k}\sigma_{ik}. \qquad (8.A.3)$$

This is purely a formal operation, but it turns out to be easier to discuss the tensor σ_{ik} than the force itself. In any case, if we can determine what the tensor σ_{ik} is, the viscous force can be derived immediately. We shall see later (Chapter 12) that σ_{ik} is one example of a stress tensor.

To understand the physical significance of the tensor σ_{ik}, consider a mass of fluid of volume V and surface S (see Fig. 8.1).

The total force per unit mass acting on a volume element is just

$$F_i^T - -\frac{\partial}{\partial x_k}(\Pi_{ik} - \sigma_{ik}), \qquad (8.A.4)$$

so that, if no outside forces are acting on the fluid, the total force acting on the fluid is just

$$F_i^T = \int_S F_i^T \, dV = -\int_V \frac{\partial}{\partial x_k}(\Pi_{ik} + \sigma_{ik}) \, dV$$

$$= -\int_S (\Pi_{ik} - \sigma_{ik}) \, dS_k, \qquad (8.A.5)$$

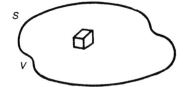

Fig. 8.1.   A fluid element in a generalized volume enclosed in a surface S.

where the last step, as in Eq. (1.C.12), follows from Gauss' law. The terms in the surface integral should look familiar. In Chapter 1, we saw that the term

$$- \int_S \Pi_{ik} \, dS_k \qquad (8.A.6)$$

could be interpreted as the sum of the pressure forces acting across the surface of the fluid and the momentum carried across the surface by the fluid motion. The additional term which we now have added,

$$\int \sigma_{ik} \, dS_k, \qquad (8.A.7)$$

has a similar interpretation. It is clearly just the force exerted across the surface S by the viscous forces which act in the fluid. In microscopic terms, it represents the momentum transferred across the surface S by inelastic collisions of the atoms near the surface.

We can learn a great deal about the tensor σ_{ik} if we ask ourselves the question "Under what circumstances will we expect no viscous forces to be present?" Clearly, from our previous descriptions, we expect the viscous forces to be absent whenever the fluid is moving in such a way that there is no relative velocity between different parts of the fluid, since then there would be no net gain or loss of energy by any part of the fluid due to inelastic atomic collisions. This situation can arise in two ways:

(i) the fluid is moving everywhere with the same velocity $\mathbf{u}$;
(ii) the fluid is in a state of uniform rotation, so that

$$\mathbf{u} = \boldsymbol{\omega} \times \mathbf{r}, \qquad (8.A.8)$$

where ω is the rotational frequency.

From the absence of viscosity in the first case, we conclude that the viscous force, and hence the tensor σ_{ik} cannot depend on the velocity itself, but must depend on the velocity through terms like $\partial u_i / \partial x_k$ and $\partial^2 u_i / \partial x_k \partial x_j \ldots$ which vanish if the velocity is a constant.

From the second case, we conclude that the tensor must vanish if $\mathbf{u} = \boldsymbol{\omega} \times \mathbf{r}$. The only combinations of derivatives of the velocity which satisfy these two conditions are

$$\frac{\partial u_i}{\partial x_k} + \frac{\partial u_k}{\partial x_i}$$

and

$$\delta_{ik} \frac{\partial u_l}{\partial x_l},$$

and, of course, a large number of terms involving second and higher derivatives of the velocity. We have no reason to expect that such terms will not be present in σ_{ik}, but it is clear that our theory would be much simpler if the viscous forces depended only on the first derivatives of the velocity. Therefore, following the lead of William of Occam,† we will assume that we are entitled to use the simplest possible theory we can write down (consistent with the conditions (i) and (ii), of course) until we are forced to do otherwise by the data. In fact, it has been found that the simple theory, in which the viscous force is assumed to depend only on the first derivatives of the velocity, is a perfectly adequate description of the motion of fluids. An alternate derivation of this result is given in Problem 12.7 in terms of the stress tensor.

This means that we can write the most general tensor in the form

$$\sigma_{ik} = \eta' \left(\frac{\partial v_i}{\partial x_k} + \frac{\partial v_k}{\partial x_i} \right) + \xi' \delta_{ik} \frac{\partial v_l}{\partial x_l}$$

$$= \eta \left(\frac{\partial v_i}{\partial x_k} + \frac{\partial v_k}{\partial x_i} - \frac{2}{3} \delta_{ik} \frac{\partial v_l}{\partial x_l} \right) + \xi \delta_{ik} \frac{\partial v_l}{\partial x_l}, \qquad (8.A.9)$$

where the coefficients η and ξ are called coefficients of viscosity. We have written σ_{ik} in the second form because this is the way it is usually found discussed in textbooks.

It should be noted in passing that by writing the most general form of σ_{ik} in Eq. (8.A.9), we have, in fact, assumed that the coefficients of viscosity do not depend on position in the fluid, and hence are really neglecting things like a possible dependence of the coefficients on temperature or other parameters in the fluid. This will be a good approximation for the applications which we wish to make, but it must be borne in mind that it may not be valid in every problem.

In most of the work which we have done up to this point, we have confined our attention to incompressible fluids; i.e. fluids for whi h the equation

$$\nabla \cdot \mathbf{v} = \frac{\partial v_l}{\partial x_l} = 0 \qquad (8.A.10)$$

is valid. We argued that this is a good approximation for liquids, but perhaps not so good for gases. For the case of incompressible fluids, the

---

†William of Occam (or Ockham), 1280–1349. He was an Oxford philosopher who had a rather exciting life, including a trial by the Pope at Avignon for heresy. He put forward the philosophical dictum "pluritas non est ponenda sine necessitate", or "multiplicity is not to be posited without necessity," which is usually known as Occam's razor. It is frequently cited in cases such as this when there is no inescapable reason to neglect complications.

viscous force becomes

$$F_i = \frac{\partial}{\partial x_k} \sigma_{ik} = \frac{\partial}{\partial x_k} \left[\eta \left(\frac{\partial v_i}{\partial x_k} + \frac{\partial v_k}{\partial x_i} \right) \right]$$

$$= \eta \frac{\partial^2 v_i}{\partial x_k \partial x_k}, \tag{8.A.11}$$

so that the Euler equation is

$$\rho \frac{\partial}{\partial t} v_i = -\frac{\partial}{\partial x_k} \pi_{ik} + \eta \frac{\partial^2 v_i}{\partial x_k \partial x_k}, \tag{8.A.12}$$

which, in a more familiar vector form becomes

$$\rho \frac{D\mathbf{v}}{Dt} = -\nabla P + \eta \nabla^2 \mathbf{v} = -\nabla P + \rho \nu \nabla^2 \mathbf{v}, \tag{8.A.13}$$

where $\nu = \eta/\rho$ is usually called the *kinematic viscosity coefficient*.

This equation is generally called the *Navier–Stokes* equation, but it will be sufficient for us to remember that it is simply Newton's second law applied to a fluid in which internal friction, or viscosity, is known to exist. The above form applies *only* to incompressible fluids. If the fluid is compressible, so that Eq. (8.A.10) is not valid, then a more complicated form of the equation could be derived (see Problem 8.1).

B. VISCOUS FLOW THROUGH A PIPE (Poisieulle Flow)

An example of viscous flow which occurs often in practical application is the flow of a fluid through a pipe. Let us consider a viscous fluid flowing through a pipe of circular cross section whose walls are perfectly rigid (later, when we consider flow of the blood in arteries, we shall consider the ramifications of allowing the walls to be elastic).

Let us further suppose that the system is in a steady state, and that the velocity of the fluid is everywhere in the z-direction (although we allow the possibility that the z-velocity may depend on the coordinate r) and that there is no dependence on the azimuthal angle (this follows from the symmetry of the problem).

The z-component of the Navier–Stokes equation then can be written

$$\rho \frac{\partial v_z}{\partial t} + \rho v_z \frac{\partial}{\partial z} v_z = -\frac{\partial P}{\partial z} + \eta \nabla^2 v_z. \tag{8.B.1}$$

Under the conditions outlined for this problem (steady state flow and the velocity being only in the z-direction and depending only on the radial coordinate), the terms on the left-hand side of the Navier–Stokes equation vanish, and we have for the z-component

$$\frac{1}{\eta}\frac{\partial P}{\partial z} = \nabla^2 v_z, \tag{8.B.2}$$

while the r-component of the equation yields

$$\frac{\partial P}{\partial r} = 0. \tag{8.B.3}$$

Equation (8.B.3), together with the requirement that the pressure not depend on the angle ϕ, implies that each plane perpendicular to the z-axis is a plane of constant pressure. Since we are dealing with an infinitely long pipe, this implies that the pressure drop in the z-direction must be uniform, or

$$\frac{\partial P}{\partial z} = -\frac{\Delta P}{\Delta l}, \tag{8.B.4}$$

where ΔP is the pressure drop in a length Δl. Equation (8.B.2) then becomes

$$\frac{1}{r}\frac{d}{dr}\left(r\frac{dv_z}{dr}\right) = -\frac{\Delta P}{\eta\,\Delta l}, \tag{8.B.5}$$

which can be integrated to give

$$v_z = \left(-\frac{\Delta P}{4\eta\,\Delta l}\right)r^2 + C_1 \ln r + C_2, \tag{8.B.6}$$

where C_1 and C_2 are constants of integration.

As with any differential equation, it is necessary to impose boundary conditions to determine these constants. In this case, we can require that the velocity be everywhere finite, including the point $r = 0$. This requirement is met by setting $C_1 = 0$.

To determine the other constant, it is necessary to specify the velocity somewhere else. We saw in treating nonviscous fluids that the boundary condition at a solid surface was that the component of velocity normal to the surface had to vanish, but that the component along the surface could be arbitrary. In the case of viscosity, however, this boundary condition does not seem adequate, since we are dealing with a fluid in which energy transfer can take place because of the existence of inelastic collisions at the atomic level.

If we think for a moment about the fluid near the wall of the tube, we will realize that the atoms in the fluid will collide with the atoms in the wall. In the idealized case where the atoms in the wall are perfectly rigid (i.e., where they can absorb an infinite amount of energy without recoiling or moving) we would expect that the atoms in the moving fluid would be reduced to a state of rest as well. In this case, then, the correct boundary condition for the fluid would be that the velocity at the surface vanish identically (and not just in the normal direction) for perfect viscosity.

Of course, in a real fluid we would expect that the fluid at the surface would have some small velocity. This phenomenon is known as "slip," and would have to be taken into account in detailed calculations. The situation is quite similar to the mechanical problem of a ball rolling across a surface. In the case of "perfect friction," we assume that the velocity of the surface of the ball at the point of contact is exactly zero. We realize, however, that in a real situation the velocity at that point will not be zero, but that some slipping will occur. Nonetheless, in most problems we are content to ignore this small effect in order to enjoy the greater simplicity of the idealized case.

Thus, in our problem, we will assume that the second boundary condition is just

$$v_z(r = R) = 0, \tag{8.B.7}$$

so that the solution for the velocity is just

$$v_z = \frac{\Delta P}{4\eta \Delta l}(R^2 - r^2), \tag{8.B.8}$$

which means that the velocity profile looks like the one shown in Fig. 8.2, i.e. the velocity is zero at the walls, and attains its maximum at the center of the pipe. Such a situation is usually referred to as Poisieulle flow.

For the sake of completeness, we note that since the total amount of fluid passing through a tube in time Δt is just

$$\Delta Q = 2\pi \Delta t \int \rho v_z r \, dr, \tag{8.B.9}$$

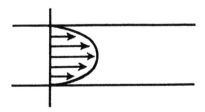

Fig. 8.2. Fully developed Poisieulle flow.

the rate of flow through a circular pipe is just

$$\frac{\Delta Q}{\Delta t} = 2\pi\rho \int_0^R \frac{\Delta P}{4\eta\,\Delta l}\,(R^2 - r^2)r\,dr$$

$$= \frac{\pi\rho}{8\eta}\left(\frac{\Delta P}{\Delta l}\right) R^4. \tag{8.B.10}$$

This result is called the Poisieulle formula. The important feature to which we shall refer when discussing blood flow is the fact that the flow rate depends on the fourth power of the radius, so that large changes in the pressure are required to compensate for small constrictions in the tube.

C.   VISCOUS REBOUND—THE VISCOSITY OF THE EARTH

One of the properties of a viscous fluid is that its response to external forces is not instantaneous. One need only think of molasses flowing from a jar to realize this. On the other hand, one would expect that the rate at which the fluid responsed to external forces would depend rather strongly on the viscosity, so that, turning the problem around, we should in principle be able to determine the viscosity of a fluid by measuring its response to known forces.

One particularly fascinating application of this idea is in measuring the viscosity of the earth. We saw in Section 2.D that in some cases, it is possible to treat the earth as a uniform fluid. If this is so, then it should be possible to make measurements which would allow us to ascribe a viscosity to that fluid. If the conjecture in the previous paragraph is correct, then we should be able to determine the viscosity of the earth by applying a known force to the surface, and then measuring the time response to that force.

Of course, we cannot produce man-made forces of sufficient magnitude to produce appreciable deformations of the earth's crust over large distances. However, nature herself has provided these forces in many cases. We shall consider two cases, which result from different geology, but obey the same physical principles.

Consider a case where there is a great weight impressed on the surface of the earth over a long period of time. Examples of this might be the existence of a lake or glacier. The surface of the earth will then be deformed by the presence of this added weight (see Fig. 8.3.). Now suppose that for some reason, the overburden is removed. This might result from the evaporation or draining of the lake, or from the melting of the glacier.

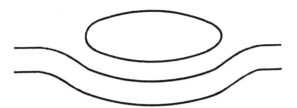

Fig. 8.3. The deformation of the earth's surface due to a glacier.

Then an imbalance of forces will exist, and the surface of the earth will slowly rebound to its original shape. The rate of rebound will depend on the viscosity. It is this sort of process that we wish to consider in this section.

The two examples which we have in mind are the so-called Fenno-Scandian uplift and Lake Bonneville, Utah. The former is the result of the removal of the glacier which covered the Scandinavian penninsula during the last ice age. Over recorded history, the level of land in this area has risen by hundreds of meters! (we will discuss actual numbers later). Lake Bonneville was, during the Pleistocene era, a large body of water which was drained and evaporated, allowing the earth there to rebound as well.

In our considerations here we will treat the earth as if it were an ordinary fluid, although it would be highly viscous. However, it should be noted that the crust of the earth (as opposed to its interior) is not anything like the surface of a fluid, but is a solid and as such can exert restoring forces of its own. We have not yet discussed the problem of solids, but when we do we shall show that although the forces generated by the elastic properties of the crust are present in both cases of interest here, they are completely negligible compared to the fluid forces which we shall assume characterize the interior of the earth.

To attack this problem, let us consider the following configuration: Let the crust of the earth in equilibrium be the plane $y = 0$, and suppose that the initial deformation of the crust is of the form

$$y_{\text{crust}}(t = 0) = \xi_0. \tag{8.C.1}$$

We call the position of the crust at any time $\xi(t)$ in order to distinguish it from the general coordinate y. To determine the position at some later time, we will, in general, have to solve the Navier–Stokes equation for this set of initial conditions. However, we recall from Chapter 4 that the easiest method of solving these equations is simply to guess at the form of a solution, and then verify that the guess does indeed work. Since we are

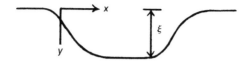

Fig. 8.4. The coordinates for the viscous rebound problem.

dealing with a highly overdamped system (the viscosity of the earth is, after all, expected to be very high), a reasonable guess for the shape of the deformation at some later time t would be

$$\xi(t) = \xi_0 e^{-kt}, \qquad (8.C.2)$$

where k is a time constant which must be determined, but which presumably depends on the viscosity. In order to relate the variable $\xi(t)$ to quantities which occur in the Navier–Stokes equation, we note that the quantity $\partial \xi / \partial t$, the rate at which the surface rises, must be the same as the y-component of the velocity of a fluid particle in the surface. We have seen this condition before in Chapter 4, where it was used to obtain the equations governing surface waves. If we make the usual approximation that the deformation is small, so that we can write $v_y(y) \approx v_y(0)$ as we did in Chapter 4, then

$$\frac{\partial \xi}{\partial t} = v_y(0). \qquad (8.C.3)$$

But since

$$\xi \frac{\partial \xi}{\partial t} = -k, \qquad (8.C.4)$$

we know that to determine the time constant k (which is, of course, a measurable quantity), we need only determine the quantity v_y from the Navier–Stokes equation.

Since the velocities in the problem are small, we can neglect the $\mathbf{v} \cdot \nabla \mathbf{v}$ term, and write

$$\rho \frac{\partial \mathbf{v}}{\partial t} = -\nabla P + \eta \nabla^2 \mathbf{v} + \rho g \hat{y}. \qquad (8.C.5)$$

Let us also write, for simplicity,

$$P' = P - \rho g y, \qquad (8.C.6)$$

so that the equation can be put into the form

$$\rho \frac{\partial \mathbf{v}}{\partial t} = -\nabla P' + \eta \nabla^2 \mathbf{v}. \qquad (8.C.7)$$

This trick, of writing terms on the right-hand side of the Navier–Stokes equation as gradients, and then incorporating them into a P' term, is one which we will use repeatedly later.

Following our standard procedure, we will now guess at the form of the solution for the quantities which appear in the Navier–Stokes equation, and in the equation of continuity. We have already guessed at the time dependence of the terms in Eq. (8.C.2). Since the velocities also have to satisfy the equation of continuity for an incompressible fluid

$$\frac{\partial v_x}{\partial x} + \frac{\partial v_y}{\partial y} + \frac{\partial v_z}{\partial z} = 0, \tag{8.C.8}$$

a reasonable guess at a solution might be

$$\begin{pmatrix} v_x \\ v_y \\ v_z \\ P' \end{pmatrix} = \begin{pmatrix} lA(y) \sin lx & \cos mz \\ B(y) \cos lx & \cos mz \\ mA(y) \cos lx & \sin mz \\ P'(y) \cos lx & \cos mz \end{pmatrix} e^{-kt}. \tag{8.C.9}$$

The assumption that the x- and z-components of the velocity have essentially the same form follows from the symmetry of the problem, and need not be considered an extra restriction on the solution. We now have to determine $B(y)$. If we put our assumed forms for the velocity into the equation of continuity, we find that

$$A(y) = -\frac{1}{l^2 + m^2} \frac{\partial B}{\partial y} = -\frac{1}{\alpha^2} \frac{\partial B}{\partial y}, \tag{8.C.10}$$

where we have made the useful definition

$$\alpha^2 = l^2 + m^2. \tag{8.C.11}$$

With this result, we will go back to the Navier–Stokes equations and eliminate the variable P' between the y- and x-components of the equation, leaving an equation for B which we can then solve.

The x-component of the Navier–Stokes equation is, when the assumed forms of the solution are inserted and the obvious cancellations made,

$$-k\rho A(y) = P'(y) + \eta\left[-\alpha^2 A(y) + \frac{d^2 A}{dy^2}\right], \tag{8.C.12}$$

which, with the aid of Eq. (8.C.10) can be written

$$-P'(y) = \left[\eta - \frac{k\rho}{\alpha^2}\right]\frac{dB}{dy} - \frac{\eta}{\alpha^2}\frac{d^3 B}{dy^3}. \tag{8.C.13}$$

The y-component of the Navier–Stokes equation, with similar substitutions and cancellations is just

$$- k\rho B(y) = -\frac{\partial P'}{\partial y} + \eta \left[-\alpha^2 B(y) + \frac{d^2 B}{dy^2} \right], \qquad (8.\text{C}.14)$$

which, if we substitute $P'(y)$ from Eq. (8.C.13), becomes

$$\frac{d^4 B}{dy^4} - \left(2\alpha^2 - \frac{k\rho}{\eta} \right) \frac{d^2 B}{dy^2} + \alpha^2 \left(\alpha^2 - \frac{k\rho}{\eta} \right) B = 0. \qquad (8.\text{C}.15)$$

In Problem 8.2, it is shown that for a large area phenomenon like the Fenno-Scandian uplift, the values of the physical constants in Eq. (8.C.15) are such that

$$\frac{\rho k}{\eta} \ll \alpha^2. \qquad (8.\text{C}.16)$$

This conclusion, of course, depends on the values of k and η which we will derive as a result of this discussion. Therefore, we will regard this approximation as a guess which we make now, and will verify after the solution of the problem has been obtained. If we make the approximation, then the solution to Eq. (8.C.15) is just

$$B(y) = - H(Gy + 1)e^{-\alpha y} + C(Dy + 1)e^{\alpha y}. \qquad (8.\text{C}.17)$$

(This can be verified by substituting the assumed form of the solution back into Eq. (8.C.15). The solution to the equation is derived in standard books on ordinary differential equations.)

There are four unknown constants in this solution, because we started with a fourth-order differential equation. We remark in passing that fourth-order equations often occur in problems involving viscosity because of the $\nabla^2 \mathbf{v}$ term in the Navier–Stokes equation which was not present in Chapter 4 (for example) when a problem similar to this was discussed for surface waves.

Since the velocity must be finite when y approaches infinity, we can immediately write

$$C = 0. \qquad (8.\text{C}.18)$$

The determination of the other constants is somewhat more complicated, and will require a little discussion of what boundary conditions are appropriate at the free surface of a viscous fluid. In Chapter 2, when we discussed the surface condition for a nonviscous static fluid, we saw that in that case the surface had to be at a constant pressure, since otherwise forces would exist at the surface which would cause elements of fluid at the surface to move, distorting the surface.

The condition that a surface be free, then, is simply that no forces act on it in such a way as to cause it to change. In the case in which we are interested at the moment, there is clearly an imbalance of forces in the y-direction, since the surface is moving in that direction. However, there should be no forces acting along the surface, in the x- or z-direction. Such forces are called shear forces. In a nonviscous fluid, no such forces exist for small velocities, but we saw in Eq. (8.A.5) that the presence of viscosity introduces a new force, depending on σ_{ik}. The condition that no shear forces exist at the surface $y = 0$ must then be that

$$\sigma_{yx}(y = 0) = \sigma_{yz}(y = 0) = 0. \qquad (8.C.19)$$

Because of the symmetry, we will consider only one of these conditions. By definition

$$\sigma_{yx} = \eta\left[\frac{\partial v_y}{\partial x} + \frac{\partial v_x}{\partial y}\right], \qquad (8.C.20)$$

which, using our assumed solutions becomes

$$\sigma_{yx} = \eta l \sin lx \cos mz\, e^{-kt}\left[B(y) + \frac{dA}{dy}(y)\right], \qquad (8.C.21)$$

so that imposing the boundary condition, we find

$$B(y = 0) = \frac{dA(y = 0)}{dy} = -\frac{1}{\alpha^2}\frac{d^2 B(y = 0)}{dy^2}, \qquad (8.C.22)$$

where the second equality follows from Eq. (8.C.10). This implies

$$G = \alpha. \qquad (8.C.23)$$

Thus, we find that the y-component of the velocity of the fluid is just

$$v_y = - H(\alpha y + 1)e^{-\alpha y} \cos lx \cos mz\, e^{-kt}. \qquad (8.C.24)$$

It remains to find a relationship between v_y and the displacement ξ. One such relation has, of course, been obtained in Eq. (8.C.4). There is another which can be obtained if we look at the forces in the y-direction at the surface. From the above discussion about the forces exerted by viscosity at the surface, it is clear that the force in the y-direction at the surface must be

$$F_y = - P' + 2\eta\frac{\partial v_y}{\partial y}\bigg]_{y=0} = - P' + \sigma_{yy}(y = 0)$$
$$= 2\eta\alpha H \cos lx \cos mz\, e^{-kt}, \qquad (8.C.25)$$

where we have used Eqs. (8.C.13), (8.C.17), and (8.C.23) to evaluate P' at $y = 0$.

From the principle of Archimedes, this must be the buoyant force, and must therefore be equal to the weight of the displaced liquid. From Fig. (8.4), the weight of displaced liquid at any point is just given by

$$F_B = \rho g \xi, \tag{8.C.26}$$

so that equating Eq. (8.C.26) to Eq. (8.C.25) gives

$$\xi = \frac{2\eta\alpha}{\rho g} H \cos lx \cos mz \, e^{-kt}. \tag{8.C.27}$$

This expression, which gives the displacement of the surface in terms of the viscosity of the earth, is precisely the expression which we seek. If we use Eqs. (8.C.3), (8.C.4), (8.C.24), and, (8.C.27), we find that the time constant for the rebound is given by

$$k = \frac{\rho g}{2\eta\alpha}. \tag{8.C.28}$$

Thus, by measuring the rate at which the rebound of the earth's crust is proceeding with time, the viscosity of the earth can be estimated. Actually, in a populated area like Scandinavia, this is not as complicated as it sounds, since one can look at old wharves which are now far inland, or at geological evidence. A full discussion of the measurements in the case is given in the text by Heiskanen and Vening Meinesz (1958). For our example, we note that in Scandinavia, the deflection in 8000 B.C. was 556 m, and is about 80 m today, so that the time constant is just

$$k \approx 6 \times 10^{-12} \sec^{-1},$$

which leads to a viscosity estimate of

$$\eta \approx 10^{22} \text{ poise}.$$

For Lake Bonneville, however, the upward deflection is estimated to be about 64 m in 4000 years. This leads to an estimated viscosity of

$$\eta \approx 10^{21} \text{ poise}.$$

The differences between these two could be due to a number of causes. In our development, we have assumed that the overburden was lifted instantaneously, whereas in both cases we considered—the melting of a glacier and the emptying of a lake—the removal of the overburden would take place over a time scale which is not terribly small compared to that of the rebound. We have also neglected the fact that the earth is not a perfect fluid, but in fact changes density appreciably over distances of the order of magnitude which we are considering here. Finally, the differences might simply be a result of the fact that the earth is just a little less rigid in North America than it is in Northern Europe.

But whatever the outcome of the discussion of the details of this type of analysis, the important point for our discussion is that it is possible, starting with the simple Navier–Stokes equation, to look at the process of elastic rebound in the crust of the earth and come up with reasonable estimates of the earth's viscosity. This illustrates again the point which was made in the first chapter—that given a few simple physical principles which govern the behavior of fluids, there is almost no end to the number of interesting examples which can be described with them.

SUMMARY

We have seen that the effects of viscosity can be included in our description of fluids by the addition of a term to the Euler equation. For incompressible fluids, this term is of the form $\eta \Delta^2 v$, where η is called the coefficient of viscosity. The examples of the flow of a fluid through a rigid pipe and the viscous rebound of the earth's surface after the removal of an overburden like a glacier were worked out.

PROBLEMS

8.1. Derive the form of the Navier–Stokes equation for the case of a compressible fluid whose coefficients of viscosity are constant.

8.2. Verify that for the Fenno-Scandian uplift area, which is approximately 1400 km on a side, the approximation

$$\alpha^2 \gg \frac{\rho k}{\eta}$$

is valid.

8.3. Show that the boundary condition in Eq. (8.C.10) and the subsequent determination of the coefficient in Eq. (8.C.23) imply that there is no motion of the fluid in the x- or z-direction in the case of viscous rebound. Is this consistent with the boundary conditions we have imposed on the problem?

8.4. The introduction of viscosity means that there is a new mechanism for dissipating energy in a fluid system. Let us repeat the energy balance analysis of Section 1.E for an incompressible viscous fluid.

(a) Show that the Navier–Stokes equation leads to the result

$$\frac{\partial}{\partial t}\left(\frac{1}{2}\rho v^2\right) = -\frac{\partial}{\partial x_i}\left[\rho v_i\left(\frac{1}{2}v^2 + \frac{P}{\rho}\right) - v_i\sigma_{ik}\right] - \sigma_{ik}\frac{\partial v_i}{\partial x_k}.$$

(b) Hence show that

$$\frac{\partial\left(\frac{1}{2}\rho v^2\right)}{\partial t} = \int_S \left[\sum \rho v_i\left(\frac{1}{2}v^2 + \frac{P}{\rho}\right) - v_i\sigma_{ik}\right] dS_k - \int_V \sigma_{ik}\frac{\partial v_i}{\partial x_k}\, dV.$$

(c) Using the definition of σ_{ik}, show that an appropriate choice of the surface S leads to

$$\frac{\partial\left(\frac{1}{2}\rho v^2\right)}{\partial t} = \frac{\partial T}{\partial t} = -\frac{\eta}{2}\int_V \left(\frac{\partial v_i}{\partial x_k} + \frac{\partial v_k}{\partial x_i}\right)^2 dV,$$

(d) and hence

$$\frac{\partial T}{\partial t} = -\eta\int_S \frac{\partial}{\partial x_i} v^2\, dS_i.$$

8.5. Consider a fluid of viscosity η flowing between two infinitive parallel plates a distance h apart. Let there be a pressure gradient dP/dz exerted by some outside agency, so that the fluid will flow in the z-direction.

(a) Calculate the velocity profile of the fluid between the plates.

(b) Hence calculate the tensor σ_{ik} in the fluid.

(c) Show that there will be a force in the z-direction per unit area on each plate given by

$$F_z = -\frac{1}{2}h\frac{dP}{dz}.$$

This phenomenon, in which a viscous liquid exerts a force on the material at its boundary, is called *drag*.

8.6. Repeat Problem 8.5 for the case where the upper plate is moving in the z-direction with velocity **V**.

8.7. The general method outlined in the above two problems can be applied to calculating the drag on any body moving through a fluid (or, equivalently, a stationary body around which a fluid flows). One case which can be solved explicitly is that of a sphere in a fluid. The result of this calculation, called *Stoke's formula*, says that the drag force on a sphere of radius a in a fluid which is moving with velocity V relative to the sphere, is given by

$$F = 6\pi R\eta V.$$

Derive this result by calculating the velocity field around a sphere, deriving σ_{ik} from the field, and integrating over the sphere to find the force. (*Hint*: You may want to consult some of the texts cited in Chapter 1, since the derivation is somewhat complicated.)

8.8. Stoke's formula tells us what the effect of air resistance would be on a falling sphere. There is a common folktale involving Galileo which says that he discovered that the acceleration due to gravity was independent of the mass by dropping different weights off of the leaning tower of Pisa. Calculate the difference in arrival times between two spheres whose masses are a factor of q different, but whose radii are the same, if they are dropped from rest from a height h. Use this result to comment on the historical validity of Galileo's experiment.

8.9. Consider the flow of a fluid in a two-dimensional plane. Let us define

$$\omega = \nabla \times \mathbf{v})_z.$$

Show that the Navier–Stokes equation and the equation of continuity imply that

$$\frac{D\omega}{Dt} = \nu \nabla^2 \omega.$$

The variable ω is usually called the *vorticity*, and this equation is called the *vorticity transport equation*. Does it resemble any other equation you know of?

8.10. Show that the potential flow of an incompressible fluid will automatically satisfy the Navier–Stokes equation provided that it satisfies the corresponding Euler equation.

8.11. Consider two cylinders of radii r_1 and r_2, rotating at angular speed ω_1 and ω_2, respectively.

(a) Write down the Navier–Stokes equation and the boundary conditions which must apply in this case.

(b) Show that if the inner cylinder is held fixed, the torque per unit length exerted by the outer cylinder is

$$M = 4\pi\eta \frac{r_1^2 r_2^2}{r_2^2 - r_1^2} \omega_2.$$

This result has been utilized as a means of measuring the viscosity of fluids.

(d) Show that in the case of a single cylinder rotating alone,

$$u(r) = \frac{r_1^2 \omega}{r}.$$

(*Hint*: This is a limit of the result in part (b).)

8.12. Consider a flat plate which is initially at rest in an infinite fluid, and which at time $t = 0$ is instantaneously accelerated to its final velocity V, which we will take to be along the plate.

(a) Show that the Navier–Stokes equation reduces to

$$\frac{\partial u}{\partial t} = \nu \frac{\partial^2 u}{\partial x^2},$$

where x is the coordinate perpendicular to the plate.

(b) If we assume a solution of the form

$$u = vf(\xi),$$

and define

$$\xi = \frac{x}{2\sqrt{\nu t}},$$

show that f is determined by

$$\frac{d^2f}{d\xi^2} + 2\xi\frac{df}{d\xi} = 0.$$

(c) Solve this equation (*Hint*: Look up the incomplete error function), and sketch the velocity near the wall.

8.13. Work through Problem 8.12 for the case where the wall is oscillating, so that its velocity is

$$V = v_0 \cos \omega t.$$

Show that the fluid velocity is given by

$$u(x, t) = V_0 e^{-\sqrt{\omega/2\nu}\,x} \cos\left(\omega t - x\sqrt{\frac{\omega}{2\nu}}\right).$$

These two examples are called *Stoke's first and second problems*.

8.14. In both of the above two problems, the fluid at large distances was essentially at rest, while the fluid near the moving plate was in motion. Show that the distance to the point at which the velocity has been reduced to about 1% of the velocity of the plate is given in both cases by

$$\delta \approx \sqrt{\nu t},$$

where t is a typical time in the problem. The significance of this result will become clear in the next chapter.

REFERENCES

In addition to the texts cited in Chapter 1, an excellent reference on the topic of viscosity is

H. Schlichting, *Boundary Layer Theory*, McGraw-Hill, New York, 1968.
This is a very thorough and surprisingly readable account of the theory of the flow of viscous fluids, leading to very good discussions of aerodynamics and turbulence.

For a discussion of the Fenno-Scandian problem, see

W. A. Heiskanen and F. A. Vening Meinesz, *The Earth and Its Gravity Field*, McGraw-Hill, New York, 1958.

9

The Flow of Viscous Fluids

Things are seldom what they seem.

GILBERT AND SULLIVAN
HMS Pinafore

A. THE REYNOLDS NUMBER

In the previous section, we examined the problem of the flow of a viscous fluid in a pipe, the Poisieulle problem. This is one of the simplest examples of the steady-state flow of fluids whose viscosity cannot be neglected. There are several important conclusions which can be drawn from this calculation. In the first place, the viscous boundary condition, which states that the fluid must be at rest at a rigid boundary, gives rise to flow patterns which are quite different from what we would expect in a nonviscous fluid, where only the normal component of the velocity must vanish.

In the second place, in fully developed viscous flow, the nonlinear terms in the Navier–Stokes equation cannot, in general, be ignored. This means that except for very simple geometries, like a circular pipe, the equations themselves will be nonlinear, and therefore quite difficult to solve. We shall see this in the example in the next section.

Before going on, however, we will study one general property of the Navier–Stokes equation which is extremely important in applications. For steady-state flow, we have

$$\frac{Dv}{Dt} = (\mathbf{v} \cdot \nabla)\mathbf{v} = -\frac{1}{\rho}\nabla P + \nu \nabla^2 \mathbf{v}. \tag{9.A.1}$$

Now suppose that we have a system in which V, L, and P are "typical" velocities, lengths, and pressures. For example, in the case of Poisieulle flow, a typical velocity might be

$$V = C(r = 0), \tag{9.A.2}$$

the maximum of the velocity profile, while a typical length might be given by

$$L = a,$$

and a typical pressure by

$$P_1 = P(z = 0).$$

In any problem, such typical values of the parameters can be defined.
Now let us change variables, so that

$$\omega = \frac{\mathbf{v}}{V},$$

$$\mathbf{x} = \frac{\mathbf{x}}{L}, \tag{9.A.3}$$

and

$$P = \frac{P}{P_1}.$$

If we insert these new variables into Eq. (9.A.1) and divide by V^2/L, we find a new equation

$$\frac{D\omega}{Dt} = -\frac{P_1}{\rho V^2}\nabla P + \frac{\nu}{VL}\nabla^2\omega, \tag{9.A.4}$$

which is now written entirely in terms of dimensionless quantities. The collection of variables on the right is given a special name. If we write

$$R = \frac{VL}{\nu} = \frac{VLP}{\eta}, \tag{9.A.5}$$

then R is called the *Reynolds number*.

The physical significance of the Reynolds number can best be understood by considering the forces acting on an infinitesimal volume in a fluid in steady-state flow (see Fig. 9.1). Neglecting the viscous term is equivalent to neglecting $\eta\nabla^2 V$ with respect to $(\mathbf{V}\cdot\nabla)\mathbf{V}$ in the Euler equation. This latter term is sometimes called the "inertial force", since it represents the momentum carried in the movement of the fluid. From Eq. (8.A.9) for an incompressible fluid, the viscous force in the z-direction along the bottom face of the cube is just

$$\sigma_{zy}dx\,dz$$

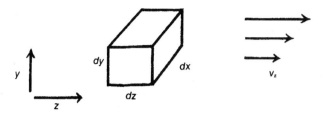

Fig. 9.1.   Forces on an infinitesimal volume element.

while that along the top face is just

$$\left(\sigma_{zy} + \frac{\partial \sigma_{zy}}{\partial y}\, dy\right) dx\, dz = \eta \frac{\partial v_z}{\partial y}\, dx\, dz + \eta \frac{\partial^2 v_z}{\partial y^2}\, dx\, dy\, dz,$$

so that the net frictional force is given by

$$\eta \frac{\partial^2 v_z}{\partial y^2}\, dx\, dy\, dz. \tag{9.A.6}$$

On the other hand, the definition of the convective derivative gives

$$\rho v_z \frac{\partial v_z}{\partial z}\, dx\, dy\, dz \tag{9.A.7}$$

for the inertial force acting on the body. If we now assume that v_z varies appreciably over a distance L, so that

$$\frac{\partial v_z}{\partial z} \approx \frac{V}{L}$$

and

$$\frac{\partial^2 v_z}{\partial y^2} = \frac{V}{L^2},$$

then the ratio of inertial to viscous forces is just

$$\rho \frac{VL}{\eta} = \frac{VL}{\nu} = R, \tag{9.A.8}$$

which is the Reynolds number. That this ratio should come up again is not surprising—the Reynolds number is the only dimensionless parameter which can be formed from the variables in the simple flow problem.

A word of caution must also be inserted at this point. The definition of the Reynolds number is somewhat arbitrary. For example, we could have chosen the diameter of the pipe instead of the radius in choosing L, the typical length in Eq. (9.A.5). This would have made a difference of a factor of two in the definition of the Reynolds number. Thus, when a Reynolds number is defined, some care should be taken in specifying exactly which lengths, pressures, and velocities are being taken as "typical," so that comparisons with other calculations (perhaps using different definitions) can be made.

Once the Navier–Stokes equation has been put into dimensionless form, as in Eq. (9.A.4), a very interesting result emerges. Suppose that we had two different situations in which two different fluids were flowing in (or around) materials which had similar shapes, but were of a different size. For example, we might be considering the flow of air around an obstruction in a large tunnel and the flow of blood around an obstruction in an artery. Suppose further that the flows were adjusted so that the ratio $P_1/\rho V^2$ were the same in each case, and so that the two flows had the same Reynolds number. Then a glance at Eq. (9.A.4) tells us that these two situations will be governed by exactly the same equation of motion. This means that except for the difference in scale, the two flows will be identical. This is called the *law of similarity*, and is of obvious usefulness in many applications of hydrodynamics. The example most familiar to the reader would be the wind tunnel, in which small-scale model airplane components can be tested. (See Problem 9.2.)

B. BOUNDARY LAYERS

We have repeatedly referred to the fact that there is a great deal of mathematical complexity in the Navier–Stokes equation. One conse quence of this in the nineteenth century was that two parallel and rather unconnected fields of study had developed in fluid mechanics. One was called theoretical hydrodynamics and involved the working out of the Euler equation for perfect fluids. We have seen in previous chapters that this sort of thing is capable of describing a large portion of the world around us. When it came to dealing with problems of flow around or through material objects, however, it was a rather dismal failure. In Chapter 8, we saw how even the simple problem of flow through a circular pipe demanded the inclusion of viscosity in the equations of motion. Consequently, the separate discipline of hydraulics grew up. This was largely an experimental engineering venture, and made little contact with the theory of fluids as we are discussing it in this text. The two disciplines were brought together in the early 1900s by Ludwig Prandtl, who developed the theory of boundary layers.

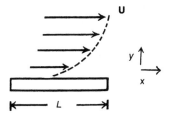

Fig. 9.2.   Flow of a viscous fluid near a plate.

Perhaps the best way to understand the idea behind boundary-layer theory is to consider the case of flow past a plate (see Fig. 9.2). Let U denote the velocity far from the plate. From our discussion of viscosity, we know that the velocity must vanish at the plate, so that there must be some variation of the velocity with y as shown.

Now if we were to form the Reynolds number for this system, our first impulse would be to take U as the typical velocity and L to be the length of the plate, to give

$$R_L = \frac{UL}{\nu} \tag{9.B.1}$$

for the Reynolds number associated with the length L.

Now in a typical situation, R_L can attain values in the hundreds or even thousands. From the reasoning in the previous section, we would conclude that for such cases, the viscous forces would be completely negligible, and we could use the Euler equation to describe the flow.

This would immediately lead to problems, however, as can be seen by considering the case of a pressure which is uniform in the y-direction. In this case, the flow of the fluid would have to be uniform as well, and hence could not vanish at the plate. This is the basic conflict between the two points of view discussed above.

Prandtl's solution was quite simple. He pointed out that while the Reynolds number defined as in Eq. (9.B.1) may be useful throughout most of the fluid, and may represent the ratio of inertial to viscous forces there, it does not do so near the plate. We can see this quickly by noting that the inertial term goes as V^2 while the viscous term goes as V, so that as V approaches zero, there must be some point at which the inertial term becomes less than the viscous term, even though in the main body of the fluid it is much larger. Thus, there will be some small region near the plate where viscous forces will dominate the motion, even though they can be neglected everywhere else. This small region is called the *boundary layer*.

Let us put the intuitive reasoning in the above paragraph into more precise form. The equation describing the flow in Fig. 9.2 is just

$$U\frac{\partial U}{\partial x} = -\frac{1}{\rho}\frac{\partial P}{\partial x} + \nu\left[\frac{\partial^2 U}{\partial x^2} + \frac{\partial^2 U}{\partial y^2}\right], \tag{9.B.2}$$

where U is a velocity in the y-direction, but can, in general, depend on both y and x.

As before, we can write

$$\frac{\partial U}{\partial x} \sim \frac{U}{L}. \tag{9.B.3}$$

In a similar way, the derivatives in the viscous term near the plate can be written

$$\frac{\partial^2 U}{\partial y^2} \sim \frac{U}{L^2}$$

and

$$\frac{\partial^2 U}{\partial x^2} \sim \frac{U}{\delta^2}, \tag{9.B.4}$$

where the second expression is true only in the region in which the velocity is making its rapid transition from zero to U. This, of course, is the region near the plate, and its thickness we denote by δ. Clearly, the viscous and inertial terms will be comparable when

$$\frac{U^2}{L} \sim \frac{\nu U}{\delta^2} \tag{9.B.5}$$

or, using Eq. (9.B.1), when

$$\frac{\delta}{L} \sim \sqrt{\frac{\nu}{UL}} = \frac{1}{\sqrt{R_L}}. \tag{9.B.6}$$

Thus, we see that there is a small region, whose extent varies inversely with the square root of the Reynolds number, in which the viscous forces cannot be neglected. It is precisely this region which is important when we are dealing with things like the viscous drag on an object in a moving fluid (such as an airfoil). This explains why the simple nonviscous theory could not be used in so many important applications, and why the inclusion of viscosity was necessary in the design of airfoils and similar things.

Actually, as we have defined the Reynolds number, it depends on the length L of the plate. We see from Eq. (9.B.6) that the actual size of the boundary layer can be expected to increase as $\sqrt{L}$. To proceed farther,

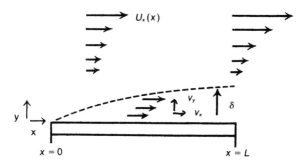

Fig. 9.3. The development of the boundary layer.

and, in particular, to determine the constant of proportionality in Eq. (9.B.5), it will be necessary to work out the equations more exactly.

Let us consider the problem of flow past a plate in more detail. Let the plate start at $x = 0$, and be of length L (see Fig. 9.3). Let the velocity profile at large y be given by $U_x(x)$, $U_y = 0$, and let the velocity in the boundary layer have components v_x and v_y which are both, in general, functions of both x and y. Let us assume that outside of the boundary layer it is reasonable to treat the flow as frictionless, and that δ, the thickness of the boundary layer, is much less than L.

The equations of motion then become

$$v_x \frac{\partial v_x}{\partial x} + v_y \frac{\partial v_x}{\partial y} = -\frac{1}{\rho}\frac{\partial P}{\partial x} + \nu\left(\frac{\partial^2 v_x}{\partial x^2} + \frac{\partial^2 v_x}{\partial y^2}\right) \qquad (9.B.7)$$

and

$$v_x \frac{\partial v_y}{\partial x} + v_y \frac{\partial v_y}{\partial y} = -\frac{1}{\rho}\frac{\partial P}{\partial y} + \nu\left(\frac{\partial^2 v_y}{\partial x^2} + \frac{\partial^2 v_y}{\partial y^2}\right), \qquad (9.B.8)$$

while continuity tells us that

$$\frac{\partial v_x}{\partial x} = -\frac{\partial v_y}{\partial y}. \qquad (9.B.9)$$

The first approximation which we shall make is that the pressure everywhere can be written as the pressure which would obtain if there were no viscosity (and which does actually exist outside of the boundary layer). One way of justifying this approximation is to use the result of Problem 9.3 that $\partial P/\partial y$ is of order δ, so that the pressure difference across the boundary layer must be

$$\Delta P_\delta = \int_0^\delta \frac{\partial P}{\partial y}\, dy \sim \delta^2 \qquad (9.B.10)$$

so that the pressure at the surface is, to order δ^2, the same as the pressure at the boundary layer. This pressure, in turn, must be the same as the pressure associated with nonviscous flow.

One important result follows immediately. From the Bernoulli equation, the pressure associated with nonviscous flow is given by

$$P + \frac{1}{2}\rho U_x^2(x) = \text{const.}$$

Since in our problem, U_x is a function of x only, we have

$$\frac{\partial P}{\partial y} = 0. \tag{9.B.11}$$

Using this result, and the results of the dimensional analysis in Problem 9.3, we find that the Eq. (9.B.7) becomes

$$v_x \frac{\partial v_x}{\partial x} + v_y \frac{\partial v_x}{\partial y} = -\frac{1}{\rho}\frac{\partial P}{\partial x} + \nu \frac{\partial^2 v_x}{\partial y^2}. \tag{9.B.12}$$

This equation and the equation of continuity, taken together, are called the *Prandtl equations*, and they describe the boundary-layer flow. They constitute two equations in two unknowns, and hence can be solved (the pressure is not an unknown here, since it is given by the flow at large y).

Consider now the case where

$$U_x(x) = U_0 = \text{const.},$$

and hence, from the Bernoulli equation,

$$\frac{\partial P}{\partial x} = 0.$$

In this case, the equations which must be solved are

$$v_x \frac{\partial v_x}{\partial x} + v_y \frac{\partial v_x}{\partial y} = \nu \frac{\partial^2 v_x}{\partial y^2} \tag{9.B.13}$$

and

$$\frac{\partial v_x}{\partial x} = -\frac{\partial v_y}{\partial y},$$

subject to the boundary conditions that

$$v_x = 0,$$
$$v_y = 0, \tag{9.B.14}$$

at $y = 0$, and

$$v_x(y \to \infty) = U_0.$$

A technique which is often useful in solving hydrodynamic equations involves the introduction of a *stream function*. Suppose that we define a function ψ by the relations

$$v_x = \frac{\partial \psi}{\partial y} \tag{9.B.15}$$

and

$$v_y = -\frac{\partial \psi}{\partial x}.$$

Then the equation of continuity in terms of the stream function is just

$$\frac{\partial^2 \psi}{\partial x\, \partial y} = \frac{\partial^2 \psi}{\partial y\, \partial x},$$

and is automatically satisfied. The Navier–Stokes equation becomes

$$\frac{\partial \psi}{\partial y} \frac{\partial^2 \psi}{\partial x\, \partial y} - \frac{\partial \psi}{\partial x} \frac{\partial^2 \psi}{\partial y^2} = \nu \frac{\partial^3 \psi}{\partial y^3}, \tag{9.B.16}$$

while the boundary conditions are

$$\psi(y = 0) = \frac{\partial \psi}{\partial x}(y = 0) = 0 \tag{9.B.17}$$

and

$$\frac{\partial \psi}{\partial x}(y \to \infty) = 0.$$

We see, then, that writing the Navier–Stokes equation in terms of stream functions, while it automatically satisfies continuity, also leads to an extremely complicated third-order nonlinear differential equation which must be solved. We can make some progress toward a solution by making a change of variables. If we let

$$\xi = y \sqrt{\frac{U_0}{\nu x}}, \tag{9.B.18}$$

and write

$$\psi = \sqrt{\nu x U_0} f(\xi), \tag{9.B.19}$$

then substitution into Eq. (9.B.13) gives a new equation in terms of the function f which is

$$f \frac{d^2 f}{d\xi^2} + 2 \frac{d^3 f}{d\xi^3} = 0, \tag{9.B.20}$$

with the boundary conditions

$$f(0) = 0,$$
$$\frac{df}{d\xi}(0) = 0, \tag{9.B.21}$$
$$\frac{df}{d\xi}(\infty) = 1.$$

This equation is solvable in principle, and can, in fact be solved numerically. A tabulation of the function f is given in the text by Schlichting mentioned in the bibliography. For our purposes, we simply

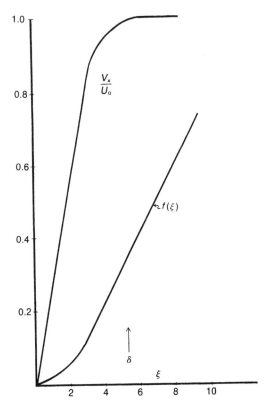

Fig. 9.4. A graph of the solutions to Eq. (9.B.20) as a function of η. The position of the boundary layer is indicated by an arrow.

plot in Fig. 9.4 the ratio

$$\frac{df}{d\xi} = \frac{v_x}{U_0} \qquad (9.B.22)$$

versus ξ. This shows the approach to asymptotic values of the velocity as we come away from the plate.

There is, of course, a certain amount of ambiguity in defining the width of the boundary layer, since the transition between zero velocity and U_0 is smooth. However, it is customary to define the edge of the boundary layer as the point at which the velocity has attained 99% of its asymptotic value. For the graph, this occurs at about $\xi = 5$, so that we have

$$\delta = 5 \sqrt{\frac{\nu x}{U_0}}, \qquad (9.B.23)$$

which agrees, of course, with the order of magnitude estimate given in Eq. (9.B.6).

An important feature of this result is that the width of the boundary layer grows as the square root of the distance along the plate, a result which was anticipated in the previous discussion. Thus, the farther we go downstream, the wider the region over which the role of viscosity is important. This makes sense in terms of the classical idea of viscosity as a frictional force between adjacent layers of fluid, since the layers near the plate will feel the forces first, and pass them along to the next layer.

One property of boundary-layer theory, which is quite important in applications but which we shall not discuss in detail here, occurs when the flow of the fluid inside of the boundary layer is in the reverse direction from the flow outside. This is called *separation*, and gives rise to eddies in the flow, and is one important aspect of the transition from laminar to turbulent flow.

SUMMARY

If the Navier–Stokes equation is put into dimensionless form, the Reynolds number can be defined and represents a measure of the relative importance of frictional and inertial forces in the fluid. While viscous forces may be small throughout the fluid taken as a whole, there is a small region near a stationary body, called the boundary layer, where they cannot be neglected. An example of viscous flow past a plane sheet was worked out to illustrate the transition from the boundary layer to the main body of the fluid.

PROBLEMS

9.1. Consider Poisieulle flow in a tube to be a model of the flow of the blood in an artery. A typical size for an artery would be 1 cm, and a typical pressure would be 100 mm of mercury.

(a) Calculate the Reynolds number for this type of flow, given that a typical value of viscosity for blood is three or four times that of water.

(b) How fast must the velocity of the blood be in order to allow us to drop the viscous term in the Navier–Stokes equation?

(c) How fast must it be to allow us to drop the nonlinear term?

9.2. Suppose that we wanted to make measurements of blood flow, but for various reasons wanted to use the flow of air in a tube as a scale model of blood flow in an artery. How would you go about designing the scale model, and what pressures and flow rates would you use in the experiment?

9.3. (a) Using the techniques of Section 9.A, put Eqs. (9.B.7) and (9.B.8) into dimensionless form.

(b) Show by dimensional analysis of the y-component of the Navier–Stokes equation that

$$\frac{\partial P}{\partial y} \sim \delta.$$

(c) Show by dimensional analysis that

$$\frac{\partial^2 v_x}{\partial x^2} \ll \frac{\partial^2 v_x}{\partial y^2}.$$

9.4. Consider the problem of flow past a plate. The drag force on the plate is given [see Eq. (8.A.9)] by

$$\psi = \eta \frac{\partial U}{\partial y}\bigg]_{y=0}.$$

(a) Using simple estimates, show that

$$\psi \approx \sqrt{\frac{\rho \eta U^3}{L}}.$$

(b) If the plate has length L and width b, find an expression for the total drag force.

(c) What is the drag on a plate six feet long and three feet wide moving through the air at 30 mph?

9.5. Verify Eqs. (9.B.20) and (9.B.22).

9.6. Why did we not have to worry about boundary layers when we solved the problem of Poisieulle flow?

9.7. The argument that the law of similarity should be expected to hold was based on the assumption that we were dealing with an incompressible fluid. If this were not the case, another dimensionless number would enter the problem. From the definition of the bulk modulus of a material as

$$\frac{1}{P}\frac{\Delta v}{v} = -E,$$

and the fact that the speed of sound is

$$C^2 = \frac{E}{\rho_0},$$

show that a fluid may be regarded as incompressible provided that the *Mach number*, M, (defined as $M = v/c$) satisfies the relation

$$\frac{1}{2}M \ll 1.$$

9.8. There are many dimensionless numbers like the Reynolds number which play important roles in various fields of hydrodynamics. Look up and define the following: Taylor, Prandtl, Eckert, and Grashof numbers.

9.9. Consider one plane inclined at an angle α to another, and moving with velocity v with respect to it, and let the space between the wedges be filled with an incompressible viscous fluid.

(a) Show that for small Reynolds number, the Navier–Stokes equation is

$$\frac{dP}{dx} = \eta\frac{\partial^2 U}{\partial y^2},$$

where x is measured along the lower (stationary) plane, and y perpendicular to it.

(b) Show that the pressure in the fluid is

$$P = P_0 + \frac{6\eta v L}{h_1^{\,2} - h_2^{\,2}}\left[\frac{(h_1 - h)(h - h_2)}{h^2}\right].$$

where L is the length along the flat plane, h_1 the height of the gap at the small end, and h_2 the height at the large end, and h the height at the point x.

(c) Calculate the total pressure and the total shearing force along the bottom plane. Show that the coefficient of friction P/F is proportional to h_2/L. Since this can be made very small, the introduction of the fluid between the two moving planes greatly reduces the friction. This is the *theory of lubrication*.

9.10. Show that in the case of two-dimensional flow, the boundary-layer equations take the form

$$v_x\frac{\partial v_x}{\partial x} + v_y\frac{\partial v_y}{\partial y} = U\frac{dU}{dx} + \nu\frac{\partial^2 U}{\partial y^2},$$

where U is the velocity outside the boundary layer.

(a) Define a stream function and write down the equation describing it.

(b) Consider now the two-dimensional flow of a viscous fluid past two planes inclined at an angle α to each other with a "sink" (a place for the fluid to flow out) at the origin. Using the techniques introduced in Problems 4.10 and 4.11, show that the flow in the boundary layer of the plane must be

$$U(x) = -\frac{U_1}{x},$$

where x is the distance measured along the plane from the sink.

(c) Defining a new variable

$$\xi = y \sqrt{\frac{-U}{x\nu}},$$

and a stream function

$$\psi = -\sqrt{\nu U_1}\, f(\xi),$$

show that the equation for f is

$$f''' - f'^2 + 1 = 0.$$

(d) Solve this equation to get

$$f' = \frac{v_x}{U} = 3 \tanh^2\left(\frac{\xi}{\sqrt{2}} + \beta\right) - 2,$$

where

$$\beta = \tanh^{-1}\sqrt{\frac{2}{3}} = 1.15.$$

(e) Hence show numerically that the width of the boundary layer in this case is approximately

$$\delta \sim 3x \sqrt{\frac{\nu}{UL}}.$$

Compare this with Eq. (9.B.6).

REFERENCES

H. Schlichting (see reference in Chapter 8) presents the best discussion of the topics in this chapter, and the reader is referred to that text for further references.

10

Heat, Thermal Convection, and the Circulation of the Atmosphere

For I had done a dreadful thing
And it would work us woe
For all averred I'd killed the bird
That made the breeze to blow

SAMUEL TAYLOR COLERIDGE
Rime of the Ancient Mariner

A. THE HEAT EQUATION AND THE BOSSINESQ APPROXIMATION

Up to this point in our studies, we have paid very little attention to the properties of fluids that have to do with temperature and heat. We know that such properties exist, however, and we will study some of their consequences in this chapter.

On the atomic scale, we are used to thinking of temperature as being associated with the motions of atoms. If the atoms have a large kinetic energy, we speak of a high temperature. Similarly, we define absolute zero classically as the temperature at which the kinetic energy vanishes. Consider what would happen if we had a fluid in which one part was heated to a temperature higher than its neighbors. In the heated section, the molecules would be moving faster. In the course of their collisions with surrounding molecules, some of this energy would, on the average, be transferred to molecules which were originally moving more slowly, thereby speeding them up. Observing this, we would say that heat was being transferred from the hot to the cold region.

In many of the applications which we have treated so far, it was reasonable to neglect effects of this type. There are some effects, like thermal convection, in which the effects of temperature dependences are the most important consideration. We begin, therefore, by discussing the classical methods of dealing with heat transfer and thermal effects.

Consider two planes in a fluid an infinitesimal distance Δx apart, with a temperature gradient $\Delta \theta$ between them (i.e. the left-hand plane is at a temperature θ, and the right-hand plane at a temperature $\theta + \Delta \theta$). Then the heat flux (heat energy per unit area per unit time) which will flow between the planes is just

$$F = - K \frac{\Delta \theta}{\Delta x},$$ (10.A.1)

where K is called the coefficient of thermal conductivity. For an arbitrary surface, then, the total heat outflow is given by

$$Q = \int_s K \frac{\partial \theta}{\partial n} dS,$$ (10.A.2)

where $\partial \theta / \partial n$ is the temperature gradient normal to the surface.

If the material inside the surface has density ρ, then a change in temperature corresponds to a change in internal energy given by

$$dU = \rho c_v \, d\theta$$ (10.A.3)

where c_v is the specific heat of the material in the volume. Thus, the rate of change of internal energy inside of the volume associated with temperature changes is just

$$\int_v \rho c_v \frac{D\theta}{Dt} dV = \frac{\partial E}{\partial t}.$$ (10.A.4)

Now the conservation of energy (the first law of thermodynamics) requires that

$$\frac{\partial E}{\partial t} = - Q,$$ (10.A.5)

i.e. it states that any change in energy in the system enclosed in the surface must be balanced by a transfer of energy across the boundary.

If we use this equality, and apply Gauss' theorem to Eq. (10.A.2) to convert the surface integral to a volume integral, we find that this equation requires that

$$\int_v \left[\rho c_v \frac{D\theta}{Dt} - K \nabla^2 \theta \right] dV = 0,$$

so that the equation which governs the temperature in an arbitrary body is just

$$\frac{D\theta}{Dt} = \kappa \nabla^2 \theta,$$ (10.A.6)

where $\kappa = K/\rho c_V$ is called the *coefficient of diffusivity*. For our purposes, we shall assume that κ is a constant for a given material.

The reader's attention is called to the similarity between the argument presented above and the derivation of the equation of continuity in Section 1.C. Why should this be so?

The heat equation must now take its place, along with the Euler equation, continuity, and the equation of state, as one of the basic equations which must be solved in describing the motion of a fluid. It is natural to ask, then, how the other equations are altered by the presence of thermal effects.

Consider the Euler equation as an example. In its most general form, it can be written

$$\rho \frac{Dv_i}{Dt} = -\frac{\partial P}{\partial x_i} + \frac{\partial \sigma'_{ik}}{\partial x_k} + \rho F_i \text{ (external)}, \qquad (10.\text{A}.7)$$

where σ'_{ik} is defined in Eq. (8.A.9). In general, both of the coefficients of viscosity, ξ and η, can depend on the temperature of the fluid. More important, the viscous term in the Euler equation will reduce to the familiar $\eta \nabla^2 v$ only in the case of an incompressible fluid. If, for some reason, we cannot use the simplified form of the equation of continuity which applies to an incompressible fluid, the Euler equation will become very difficult to handle mathematically.

The general equation of continuity takes the form

$$\frac{\partial \rho}{\partial t} + \nabla \cdot (\rho \mathbf{v}) = 0. \qquad (10.\text{A}.8)$$

If ρ is constant, this will simplify to

$$\nabla \cdot \mathbf{v} = 0. \qquad (10.\text{A}.9)$$

For a fluid in which thermal effects are important, however, such a simplification is not possible, since there will be an equation of state which will link density to temperature. For an ideal gas, for example, we would have

$$P = \rho R \theta.$$

Thus in introducing thermal effects, we are in effect loosing Eq. (10.A.9). Since the relatively simple form of this equation was instrumental in allowing us to solve problems up to this point, this is a rather serious matter.

So long as we were dealing with systems like the tides, the approximation of incompressibility was good, since for the type of temperature differences which exist in that problem, density changes are not too great. This is actually a reasonable approximation for most liquids (except in the case of thermal convection). There are some situations where it is not such a good approximation, and we are faced with the problem of either treating the equations of motion in their full complexity (a formidable task), or finding another reasonable approximation scheme for discussing thermal effects in gases. This scheme was first advanced by H. Bossinesq, and bears his name. The approximation is based on two observations:

1. The coefficients of viscosity and diffusion vary slowly with temperature for most materials.
2. In treating convection, we expect the most important effects to arise from the fact that warm air is lighter than cold air—i.e. from the way in which the gas at different temperatures is affected by gravity.

These observations led Bossinesq to propose the following approximation scheme: Ignore the variation of all quantities in the equations of motion with temperature *except* insofar as they are concerned with gravitational effects. In other words, we shall include the variation of density with temperature in the term $\rho \cdot F_i(\text{ext})$ on the right-hand side of Eq. (10.A.7), but shall treat ρ as a constant in all other equations.

This immediately results in enormous simplifications. The equation of continuity reduces to the familiar form of Eq. (10.A.9). If we are dealing with a system in which the only external force is gravity (as would be the case, for example, in considering the motion of the atmosphere), the external force term in the Euler equation becomes

$$\rho F(\text{ext}) = \rho g = g\rho_0\left(1 + \frac{\Delta\rho}{\rho_0}\right) = g\rho_0 - g\rho_0\alpha\theta, \qquad (10.A.10)$$

where α is the coefficient of expansion for the gas, defined by

$$\frac{\Delta\rho}{\rho_0} = -\alpha(\theta - \theta_0), \qquad (10.A.11)$$

and θ_0 is the temperature at which the density is ρ_0.

Thus, the Euler equation for a fluid in which thermal effects are allowed reduces to

$$\frac{D\mathbf{v}}{Dt} = -\frac{1}{\rho}\nabla P + (g - g\alpha\theta)\hat{y} + \nu\nabla^2\mathbf{v}, \qquad (10.A.12)$$

where $\hat{y}$ is a unit vector in the direction of the gravitational force, and we have defined our temperature scale so that $\theta_0 = 0$.

This approximate form of the Navier–Stokes equation, together with the continuity condition, equation of state, and heat equation, then becomes the means by which we shall describe the motion of heated fluids.

B. STABILITY OF A FLUID BETWEEN TWO PLATES

As the first example of a fluid system in which thermal effects are important, let us consider the situation shown in Fig. 10.1, in which there are two rigid plates a distance h apart, with the lower one maintained at temperature θ_1 and the upper one at θ_2. Let us begin by examining the stability of such a system.

If the system is located in a gravitational field, such as that at the surface of the earth, and if $\theta_1 > \theta_2$, our intuition tells us that it should be unstable. This follows from the fact that the warm fluid at the bottom will be less dense than the cold fluid at the top, so that the gravitational energy of the system could be lowered by letting the warm air rise and the cold air fall, as indicated in Fig. 10.1. This exchange is, of course, what we normally think of as thermal convection.

Let us see if this intuitive result can be derived from the equations of motion derived in the previous section. The first step in discussing stability is, of course, to find the equilibrium point of the system. If we define Θ as the temperature at the point y for equilibrium, then by inspection we see that the equations will be satisfied if

$$\mathbf{v} = 0 \qquad\qquad (10.B.1)$$

and

$$\frac{D\Theta}{Dt} = 0 = \nabla^2\Theta = \frac{d^2\Theta}{dy^2}, \qquad\qquad (10.B.2)$$

where the last equality follows from the assumption that the system has infinite extent in the z- and x-directions. The temperature equation can be solved to give

$$\Theta = \beta y + \theta_1, \qquad\qquad (10.B.3)$$

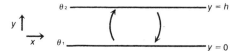

Fig. 10.1.   The geometry for the discussion of thermal convection.

where

$$\beta = \frac{\theta_2 - \theta_1}{h}.$$ (10.B.4)

This equilibrium is easy to picture—it corresponds to having the fluid completely at rest, with pressure forces balanced by gravity, and a uniform temperature gradient between the plates. In what follows, we shall investigate the stability of this system by assuming that there are small time-dependent deviations from this equilibrium, and see whether they grow as a function of time or not.

If we define a new variable by the relation

$$\theta' = \theta - \Theta = \theta - [\beta y + \theta_1],$$ (10.B.5)

where θ is the actual temperature at y, then the Euler equation in the form (10.A.12) can be written

$$\frac{D\mathbf{v}}{Dt} = -\frac{1}{\rho} \nabla(P + g\rho y) + g\alpha(\theta' + \Theta)\hat{y} + \nu\nabla^2\mathbf{v}$$
$$= -\frac{1}{\rho}\nabla\left(P + g\rho y + \gamma\rho\int^y \Theta(y')\,dy'\right) + \gamma\theta'\hat{y} + \nu\nabla^2\mathbf{v},$$ (10.B.6)

where we have written

$$\gamma = g\alpha.$$ (10.B.7)

The heat equation can also be written in terms of θ (not θ'), and is just

$$\frac{D\theta}{Dt} = \kappa\nabla^2(\Theta + \theta') = \kappa\nabla^2\theta',$$
$$= \frac{\partial}{\partial t}(\Theta + \theta') + \mathbf{v}\cdot\nabla(\theta' + \Theta).$$ (10.B.8)

Now by definition, the time derivative of Θ vanishes. Since we will be using this equation to examine departures from an equilibrium in which the velocity is zero, $\mathbf{v}$ represents departures from equilibrium, and can therefore be regarded as a small quantity. The same is true of θ'. Thus, the term $(\mathbf{v}\cdot\nabla)\theta'$ in the above can be dropped. Since Θ is a function of y only, we are left with

$$\frac{\partial\theta'}{\partial t} + \beta v_y = \kappa\nabla^2\theta',$$ (10.B.9)

where we have used the statement $\partial\Theta/\partial y = \beta$ from Eq. (10.B.3).

Let us now examine the stability of the equilibrium which we have found by looking at the behavior of the system when small perturbations from equilibrium are introduced. In particular, let us, guided by the symmetry of the problem, assume that

$$\begin{pmatrix} v_x \\ v_y \\ v_z \end{pmatrix} = \begin{pmatrix} A(y) \\ B(y) \\ C(y) \end{pmatrix} e^{ilx} e^{imz} e^{nt}. \tag{10.B.10}$$

In addition, we will see if we can satisfy all of the equations when we assume

$$\begin{pmatrix} P' \\ \theta' \end{pmatrix} = \begin{pmatrix} P'(y) \\ \theta'(y) \end{pmatrix} e^{ilx} e^{imz} e^{nt}, \tag{10.B.11}$$

where

$$P' = P + g\rho y + \gamma\rho \int^{y} \Theta(y') \, dy'.$$

The logic of this approach is as follows: We assume velocities, pressures and temperature deviation as above. We shall see that these forms can, indeed, satisfy the basic equations. We shall then look for the behavior of the system as a function of time by solving for n. If the values which we find are positive, then the system will be unstable against the type of perturbations that we have assumed. Since an arbitrary perturbation will contain some component which can be expressed as the above, this means that the system will be unstable, and will not stay in its equilibrium configuration.

The determination of the functions $A(y)$, $B(y)$, and $C(y)$ in Eq. (10.B.10) is, in general, not something which can be done by inspection, since it involves the boundary conditions at the two surfaces $y = 0$ and $y = h$. Consequently, we shall discuss the general technique which can be used to solve for these functions, but work out in detail only the simplest possible boundary conditions—that in which both planes bounding the fluid are considered to be free surfaces, so that the condition

$$v_y = 0 \qquad \text{at } y = 0, \quad y = h,$$

must hold. Clearly, if the surfaces were rigid, we would have in addition to the above the statement that not only v_y, but v_x and v_z would also vanish at these surfaces.

In addition to the boundary conditions above, we know that the velocities must satisfy the equation of continuity for an incompressible fluid, which is

$$\frac{\partial v_x}{\partial x} + \frac{\partial v_y}{\partial y} + \frac{\partial v_z}{\partial z} = 0,$$

so that aside from constants of proportionality, we must have

$$A(y) \propto C(y)$$

and

$$B(y) \propto \int^{y} A(y') \, dy'.$$

If these conditions did not hold, it would be impossible to satisfy the equation of continuity at all points in the fluid.

In general, we would assume a solution for $B(y)$ of the form

$$B(y) = e^{sy},$$

and solve for the values of s which are consistent with the boundary conditions. For the case of two free boundaries, however, we can see by inspection that

$$B(y) \propto \sin\left(\frac{q\pi}{h} y\right) = \sin sy$$

(where q is an integer) will satisfy the boundary conditions. Thus, we find that the equation of continuity and the boundary conditions will be satisfied provided that

$$\begin{aligned} A(y) &= A \cos sy, \\ B(y) &= B \sin sy, \\ C(y) &= C \cos sy, \end{aligned} \tag{10.B.12}$$

where the constants A, B, and C must satisfy the relation

$$ilA + sB + imC = 0. \tag{10.B.13}$$

If we now drop the $(\mathbf{v} \cdot \nabla)\mathbf{v}$ terms as being of second order in small quantities, we can insert the assumed forms of the solutions in Eqs. (10.B.10) and (10.B.11) into the three components of the Euler equation to obtain

$$\begin{aligned} nv_x &= -\frac{il}{\rho} P' - \nu(l^2 + m^2 + s^2)v_x, \\ nv_z &= -\frac{imP'}{\rho} - \nu(l^2 + m^2 + s^2)v_z, \\ nv_y &= -\frac{1}{\rho}\frac{\partial P'}{\partial y} + \gamma\theta' - \nu(l^2 + m^2 + s^2)v_y. \end{aligned} \tag{10.B.14}$$

If we let

$$l^2 + m^2 + s^2 = a \tag{10.B.15}$$

and

$$n' = n + \nu a, \tag{10.B.16}$$

these equations simplify to

$$n'v_x = -\frac{il}{\rho}P',$$

$$n'v_z = -\frac{im}{\rho}P',$$

$$n'v_y = -\frac{1}{\rho}\frac{\partial P'}{\partial y} + \gamma\theta'.$$

The continuity equation is just

$$ilv_x + imv_z + \frac{dv_y}{dy} = 0.$$

By differentiating the x- and z-components of the Euler equation with respect to y, we find

$$\frac{1}{\rho}\frac{\partial P'}{\partial y} = \frac{-n'}{l^2 + m^2}\frac{\partial^2 v_y}{\partial y^2},$$

which, when substituted into the y-component of the Euler equation [taking account of Eq. (10.B.12)], yields

$$n'av_y - \gamma\theta'(l^2 + m^2) = 0. \tag{10.B.17}$$

Thus, the net result of our manipulations of the Euler equation and continuity is to give us one equation relating v_y and θ'. Another such equation can be obtained by inserting our assumed forms of solution into the heat equation, giving

$$\beta v_y + [n + \kappa a]\theta' = 0. \tag{10.B.18}$$

Thus, the problem of finding a solution to the Euler equation, the heat equation and the continuity condition for perturbations of the type which we have assumed reduces to the problem of solving the above two linear equations. It is well known that solutions for v_y and θ' will exist provided that the Wronskian determinant vanishes—i.e.

$$\begin{vmatrix} \beta & n + \kappa a \\ n'a & -\gamma(l^2 + m^2) \end{vmatrix} = 0$$

or

$$\beta\gamma(l^2 + m^2) + n'na + n'\kappa a^2 = 0. \tag{10.B.19}$$

This equation determines the growth constant n. Provided this equation is satisfied, solutions of the type which we have assumed will exist. The question of stability or instability of the system reduces, then, to finding out under what conditions the value of n from the above equation will be positive.

In order to discuss the physics of this equation, let us consider a fluid which has both η and κ set equal to zero. This would correspond to an "ideal" fluid in the sense that it would be nonviscous, and the temperature of a volume element would not change, but the prime feature which we are considering, which is the change of density with temperature, is still in the problem. In this case, the condition in Eq. (10.B.19) reduces to

$$n^2 = \frac{-\beta\gamma(l^2 + m^2)}{a}. \tag{10.B.20}$$

Now there are two possibilities for n. If $\beta > 0$ (i.e. if the temperature gradient is such that the higher temperature is at the top, then $n = \pm i|n|$, and all deviations from equilibrium behave like $e^{\pm i|n|t}$—i.e. they do not grow as a function of time, but oscillate about the equilibrium configuration. Such situations are stable.

On the other hand, if $\beta < 0$, (i.e. if the lower plate is maintained at a higher temperature than the upper one), then

$$n = \pm \sqrt{\frac{|\beta|\gamma(l^2 + m^2)}{a}}, \tag{10.B.21}$$

and small deviations from equilibrium will grow exponentially in time. This, of course, is the result which we expected intuitively.

We see then that in this simple case the fluid will be unstable if there is even the smallest adverse temperature gradient. The question of what will happen to the system at large times will be discussed in the next section. But Eq. (10.B.20) already has some unfortunate consequences, since it predicts that if there is ever a situation in which air at a high level over a city is warmer than the air near the ground (this is known as an *inversion*), there will be a stable situation. When this happens in Los Angeles, as it does periodically, the effluents in the atmosphere cannot be removed by the normal circulation of the air, and a smog crisis results.

Another question we can ask at this stage is what the maximum value of n is, since this can be expected to govern the growth rate of the instabilities. From Eq. (10.B.21), it is clear that n will be a maximum when s^2 (and therefore a) is a minimum for a given l and m. Thus, the fastest growing disturbance will correspond to

$$s = \frac{\pi}{h}. \tag{10.B.22}$$

We should also note in passing that while we have been assuming that velocities will grow exponentially in time, they clearly cannot do so forever. In fact, since we have been dropping second-order terms in our derivation, which amounts to ignoring v^2 with respect to v, our solution will be invalid for large values of t in any case.

Having seen how the equations of motion can give us the result which we expected in the "ideal" case, let us now look at a more realistic fluid, in which neither ν nor κ are zero. In this case, solving Eq. (10.B.19) for n, gives

$$n = \frac{a(\kappa + \nu)}{2}\left[-1 \pm \sqrt{1 - \frac{4}{a^2(\kappa + \nu)^2}\left(\kappa\nu a^2 + \frac{\beta\gamma}{a}(l^2 + m^2)\right)}\right]. \quad (10.B.23)$$

Once more, we see that to have instability, the term under the radical must be positive, which is only possible if $\beta < 0$. Thus, an adverse temperature gradient is again a necessary (but not sufficient), condition for stability. Actually, a more precise statement of this condition is

$$|\beta|\gamma(l^2 + m^2) > \kappa\nu a^3.$$

We see that if either $\kappa = 0$ or $\nu = 0$, it will always be possible to find some $n > 0$, which is the result that we had derived previously. However, if both κ and ν are nonzero, this need not necessarily be the case.

Define

$$f(a) = |\beta|\gamma(a - s^2) - \kappa\nu a^3. \quad (10.B.24)$$

If $f(a)$ is positive, then we can have instability (see Eq. (10.B.23)), so the properties of $f(a)$ will determine stability. Let us examine this curve for fixed s as a function of $l^2 + m^2$. If $l^2 + m^2$ is very large, $f(a)$ becomes negative. Similarly, if $l^2 + m^2$ is very small, $f(a)$ becomes negative. Some possible curves for $f(a)$ are shown in Fig. 10.2. Which curve represents the

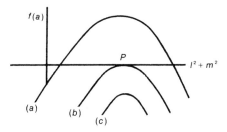

Fig. 10.2. Some sample curves of $f(a)$ as a function $l^2 + m^2$.

function depends on the choice of other parameters, including ν, κ, β, and s.

If we had a situation in which a curve such as (a) was found to hold, we note that there is a range of values of $l^2 + m^2$ for which $f(a)$ is positive, and for which n could therefore be positive, and the system could be unstable. A curve of type (c), however, would correspond to a situation where there was no choice of parameters which could satisfy the condition $n > 0$, so that no value of n would be positive. Such a curve would correspond to a system which was completely stable against all perturbations of the type which we are considering. The curve (b) corresponds to the "critical" curve at which the system moves from stability to instability. To find the value of $l^2 + m^2$ at which the system just becomes stable, we set the derivative of $f(a)$ equal to zero (this will correspond to the maximum in the curve at the point P).

$$\frac{df(a)}{da} = |\beta|\gamma - 3\kappa\nu a^2. \tag{10.B.25}$$

Since the curve (b) is specified by the equation

$$f(a) = 0$$

at P, we can solve this for the parameter $\kappa\nu$ and plug the result back into Eq. (10.B.25) to give

$$|\beta|\gamma - 3a^2\left(\frac{|\beta|\gamma(a - s^2)}{a^3}\right) = 0,$$

which corresponds to the condition

$$l^2 + m^2 = \frac{1}{2}s^2. \tag{10.B.26}$$

When this condition is satisfied, the requirement that $f(a)$ be zero at the point P becomes

$$|\beta|\gamma = \frac{27}{4}\kappa\nu s^4. \tag{10.B.27}$$

What we have derived, then, is the following: Unless $f(a)$ can be positive for some value of $l^2 + m^2$, the system will be stable against any perturbation of the type we are considering. But $f(a)$ will be positive only if

$$|\beta|\gamma > \frac{27\pi^4\kappa\nu}{4h^4}, \tag{10.B.28}$$

(since $s = q\pi/h > \pi/h$) i.e. only if the value of the function at its maximum is greater than zero. Thus, when we include viscosity in our considerations, there will be, for a given choice of material for the fluid, a range of adverse temperature gradients for which the system will be completely stable. The addition of viscosity thus makes a qualitative difference in the stability problem. The above condition is called the Rayleigh condition, after Lord Rayleigh, who first discovered its significance.

A way of visualizing this effect is to compare viscosity to static friction in mechanics. If a weight rests on a surface, it is necessary to apply some force in order to get it to move at all. There is, in addition, a critical force below which no motion will result. Viscosity plays the same role in the problem of thermal instabilities. When we apply an adverse thermal gradient, it is necessary to overcome the internal friction in the fluid in order to get it to move, and the critical value of β in the Rayleigh criterion corresponds to the critical force in the mechanical problem.

It is interesting to note in passing that every new effect which we add to the system—thermal conduction, viscosity, etc., seems to work in the direction of increasing the stability of the system against perturbations. This is a general rule, and is found to hold true for rotation and magnetic effects as well as for viscosity.

C. CONVECTION CELLS

Up to this point, we have concerned ourselves only with the question of stability of fluids in which temperature gradients exist. Since such systems are seen to be unstable, we can then ask the next question— what will the steady-state motions of the system be?

Fortunately, it is not necessary to trace through the development of instabilities as they grow in time and approach the steady-state motion. We can get the steady-state motion directly from the results of the last section by recalling we will have steady-state conditions if the time derivative of the velocity vanishes. For the type of disturbances which we treated in Eq. (10.B.10), this corresponds to getting the parameter $n = 0$ in all subsequent equations.

Thus, for the specific case of two free surfaces which was discussed in the last section, the steady-state velocities will be given by

$$v_x = A \cos sy\, e^{ilx} e^{imz},$$
$$v_y = B \sin sy\, e^{ilx} e^{imz}, \qquad (10.C.1)$$
$$v_z = C \cos sy\, e^{ilx} e^{imz},$$

where the constants A, B, and C still satisfy the auxiliary condition of Eq. (10.B.13) which is imposed by continuity. For the sake of simplicity, let us discuss the geometrically simple case where $m = 0$ and $C = 0$ (more complicated geometries will be left to the problems). This corresponds to assuming a symmetry in the z-direction, and reduces the problem of tracing out the motion of the fluid to two dimensions. The continuity condition then tells us that

$$-ilA = sB. \tag{10.C.2}$$

If we use this result and then follow the usual procedure of taking the real parts of the complex quantities in Eq. (10.C.1) to get actual physical velocities, we find

$$\text{Re } v_x = A \cos sy \cos lx,$$
$$\text{Re } v_y = \frac{l}{s} A \sin sy \sin lx. \tag{10.C.3}$$

To picture the motion associated with this velocity field, let us ploy v_y as a function of x at fixed y. It will look like the field shown in Fig. 10.3. For different choices of x, the magnitude of the velocities will be different than those pictured, but the pattern of the velocity field repeating every time we go through a distance L (which is clearly given by $L = 2\pi/l$) will reappear along every line of constant x. Thus, the fluid will naturally divide itself into regions in which periodic velocities will repeat themselves. We will refer to such units of division as cells.

We note that along the lines $x = 0$ and $x = L$, we have from Eq. (10.C.3) that

$$v_x = 0$$

and

$$\frac{\partial v_y}{\partial y} = 0.$$

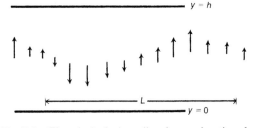

Fig. 10.3.   The velocity in the y-direction as a function of x.

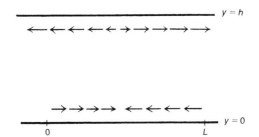

Fig. 10.4. The velocity in the x-direction as a function of y.

Thus, it follows that fluid in one cell will never leave that particular cell and enter another. The cell boundaries thus have a physical significance, in that they delineate real boundaries in the fluid, through which the fluid may not flow.

To visualize the pattern of flow inside of a given cell, let us plot v_x as a function of x for several different values of y (see Fig. 10.4).

The overall pattern of flow in the cell will then be one in which the fluid flows up from the bottom in the center of the cell, and falls down at the sides, as in Fig. 10.5. This corresponds to our general notion of convection, in which heat is transferred from the warmer to the colder to warmer surface by motion of the fluid. We should note in passing that not only does the fluid not cross the boundaries at $x = 0, L, 2L, \ldots$, but it also does not cross the boundaries at $x = L/2, 3L/2, \ldots$ The fluid in the simple case comes in "rolls," and alternate rolls involve fluid rotation in opposite directions. Two rolls together comprise what we have termed a cell.

These cells are called convection cells, or *Bénard cells*, after the French physicist who first observed them in the laboratory. They play an important part in all considerations of motion of fluids driven by thermal differences.

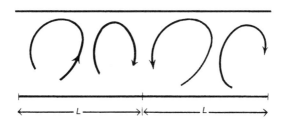

Fig. 10.5. The development of convection, or Bénard, cells.

Before leaving this topic, we note that the condition derived in the last section concerning the dimensions of the fastest growing singularity also gives us some clue as to the relationship between h, the thickness of the layer of liquid, and L, the size of the cell. We had

$$l^2 + m^2 = \frac{1}{2} s^2 = \frac{1}{2} \frac{\pi^2}{h^2},$$

recalling that $m = 0$ in our case, and the definition of L, we find

$$L = 2\sqrt{2}\, h.$$

This means that we expect the dimensions of a convection cell to be roughly the same as the depth of the liquid. This result is fairly general, although we have proved it only for the simplest possible plane geometry.

D. THE GENERAL CIRCULATION OF THE ATMOSPHERE

Probably the most important application of the theory of convection is the motions of the atmosphere due to heating effects. Bénard cells appear in many places in atmospheric motion. For example, the surface air above cities is usually warmer than the air in the surrounding countryside. This gives rise to convection cells whose scale is on the order of miles across. In a similar way, a paved shopping center (or an island in the ocean) can cause cells of somewhat smaller size. In this section, we will give a qualitative description of another type of cell—that associated with the large-scale movement of air around the earth.

Consider the temperature at the surface of the earth. The single most dominant feature of the temperature distribution is that, in simplest terms, it is warmer at the equator than at the poles. Thus, following our

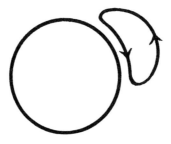

Fig. 10.6.   The Hadley cell.

discussion of Bénard cells, one might expect a net motion as pictured in Fig. 10.6—where warm air rises at the equator, and descends at the poles. This picture, called a Hadley cell, after G. Hadley, was first proposed as a way of explaining the observed wind patterns in the tropics. From a simple application of the Coriolis force, we can see that in this model, winds in the northern hemisphere would tend to blow from east to west. (It is one of the characteristics of meteorology that winds are named by the direction from which they come, rather than the direction to which they go. Winds of the type predicted by the Hadley model would thus be termed easterly winds.)

Actually, the general pattern of winds on the earth is more complicated than this. Neglecting details of local motion, the general winds patterns can be pictured as in Fig. 10.7. In the region of the tropics, from 0° to 30° north latitude, the winds are generally easterly. These are called the trade winds, and were explained by Hadley's original model. From 30° to 60° north latitude, the winds are generally westerly. This region includes most of the temperate zone of the earth. Finally, from 60° to 90° north latitudes, the winds become easterly again.

Actually, this picture is greatly oversimplified. The latitudes at which the prevailing winds change direction are not sharp dividing lines, but are smeared out, and change with the season. The structure of the region of prevailing westerlies (as we shall see later) is much more complicated than indicated in the figure. Nevertheless, for our purposes, this picture of the general circulation of the atmosphere will suffice. We note that a

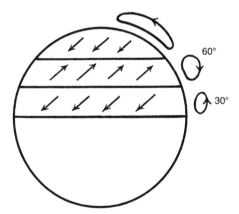

Fig. 10.7.   A simplified picture of the circulation of the atmosphere.

model in which there are three Hadley cells (see Fig. 10.7) would give the correct directions for the prevailing winds. (Like our picture of the actual wind motions, this model is greatly oversimplified.)

The transition latitudes, where the main motion of the air is in the vertical direction, are regions where there is very little wind at the surface. These regions were well known to early ocean navigators. The region at the equator is called the doldrums, and the region at about 30° north is called the "horse latitudes." The name derived from the fact that ships sailing to the New World would be becalmed when they entered this region, so that it was necessary to jettison any cargo that consumed food or water. Since these ships usually carried horses, they were the first to go. The sight of horse carcasses floating in the ocean gave the region its name.

The actual calculation of this general circulation is quite difficult, for two reasons. First, the circulation takes place in a spherical shell rather than on a plane, so the geometry is complicated, and second, the effects of the earth's rotation, as expressed in the Coriolis force, add complications to the equations of motion. Let us consider the calculation of the simple Hadley cell, without rotation, to give some flavor of what the full calculation might look like.

We begin by assuming that the temperature distribution is a function of latitude only, and ignore the temperature differences between night and day. Thus, when we define the temperature at a point, it should be regarded as the average daily temperature and not the instantaneous one. Since we are interested in solving for long-term wind patterns, this is not a drastic approximation. The short-term diurnal effects which we are neglecting can be expected to give rise to small-scale effects which, in the first approximation, do not affect the long-term winds at all. This is known as the hypothesis of *zonal heating*, and was also introduced by Hadley in 1735.

We shall once again use the Bossinesq approximation introduced in Section 10.A in which the effects of changes in density due to changes in temperature are neglected except insofar as they affect the action of the gravitational force. We shall also assume that the heating of the earth has gone on for a long time, so that the temperature distribution has stabilized and reached its steady-state value. In this case, the heat equation is simply

$$\nabla^2 \theta = 0, \qquad (10.\text{D}.1)$$

the equation of continuity is

$$\nabla \cdot \mathbf{v} = 0, \qquad (10.\text{D}.2)$$

and the Navier–Stokes equation for the steady state is just

$$0 = -\frac{1}{\rho} \nabla \left(P + \frac{1}{2} \bar{\omega}^2 \omega^2 \right) - \rho \nabla \Omega_e + \nu \nabla^2 \mathbf{v}. \tag{10.D.3}$$

In the Bossinesq approximation, the gravitational force due to the presence of the earth is just

$$\rho \nabla \Omega_e = \rho_0 (1 - \alpha \theta) G M_e \nabla \left(\frac{1}{r} \right), \tag{10.D.4}$$

where M_e is the mass of the earth (assumed to be spherical) and G is the gravitational constant.

As in the development of the simple Bénard cell, it shall be convenient to refer the temperatures and velocities to an equilibrium solution of the equations of motion. In the case of the Bénard cell, we saw that a solution existed when $\mathbf{v} = 0$. Can we find such a solution to the above equations? If we let Θ be the equilibrium temperature distribution, we see that the Navier–Stokes equation reduces in this case to

$$0 = -(1 - \alpha \Theta) G M_e \nabla \left(\frac{1}{r} \right) - \frac{1}{\rho_0} \nabla (p_1), \tag{10.D.5}$$

where p_1 is the equilibrium pressure distribution. Since $G M_e \nabla (1/r)$ is a function of r only, this equation can only be satisfied if both Θ and p_1 are functions of r alone as well. Given this, however, an equilibrium solution is indeed possible.

The physical meaning of this equilibrium solution to the equations is quite simple. It corresponds to a situation in which the atmosphere is uniformly heated (i.e. there is the same heat flow into the atmosphere at each point), and looses heat only through radiation at its upper edge. In this case, the pressure adjusts itself to balance the gravitational force.

We can now define a new temperature

$$\theta' = \theta - \Theta. \tag{10.D.6}$$

If we insert this into the gravitational force term in Eq. (10.D.4) and use the two identities,

$$\Theta \nabla \left(\frac{1}{r} \right) = -\nabla \int^r \frac{\Theta(r')}{r'^2} \, dr'$$

and

$$\theta' \nabla \left(\frac{1}{r} \right) = \nabla \left(\frac{\theta'}{r} \right) - \frac{1}{r} \nabla \theta', \tag{10.D.7}$$

the gravitational term becomes

$$\rho \nabla \Omega_e = \rho_0 GM_e \nabla \left[\frac{1 - \alpha\theta'}{r} + \alpha \int^r \frac{\Theta(r')}{r'^2} dr' \right]$$
$$- \rho_0 GM_e \alpha \frac{\nabla \theta'}{r}. \tag{10.D.8}$$

Now we saw in the derivation of the simple Bénard cell that in solving for the velocities, the pressure was eliminated between different components of the Navier–Stokes equations. Since we shall follow the same procedure here, we can split up the pressure in any way which will be mathematically convenient. In particular, we can write

$$p = p_2(1 + \delta), \tag{10.D.9}$$

where p_2, which will be defined below, is closely related to the equilibrium pressure and δ is a small parameter. Writing the pressure in this way, we can see that the gradient of the pressure which appears in Eq. (10.D.3) can be written

$$\frac{1}{\rho} \nabla p = \frac{p_0}{\rho_0} \nabla \ln p_2(1 + \delta)$$
$$= \frac{p_0}{\rho_0} \nabla \ln p_2 + \frac{p_0}{\rho_0} \nabla \delta, \tag{10.D.10}$$

where we have dropped higher-order terms in δ and made use of the identity

$$\nabla \ln p = \frac{1}{p} \nabla p.$$

The Navier–Stokes equation for the steady state can now be written using the results of Eqs. (10.D.10) and (10.D.8) as

$$0 = -\nabla \left[\frac{p_0}{\rho_0} \ln p_2 - \rho_0 GM \left(\frac{1 - \alpha\theta'}{r} + \alpha \int^r \frac{\Theta(r')}{r'^2} dr' \right) \right]$$
$$- \frac{p_0}{\rho_0} \nabla \delta + \nu \nabla^2 \mathbf{v} + GM\rho_0 \alpha \frac{\nabla \theta'}{r}. \tag{10.D.11}$$

We define p_2 such that

$$\frac{p_0}{\rho_0} \ln p_2 = \rho_0 GM_e \left(\frac{1 - \alpha\theta'}{r} + \alpha \int^r \frac{\Theta(r')}{r'^2} dr' \right) \tag{10.D.12}$$

(we see that in the case of equilibrium heating ($\theta' = 0$) p_2 becomes identical to p_1, the equilibrium pressure). With this assignment of p_2, the

Navier–Stokes equation finally takes the form

$$0 = GM\rho_0\alpha\frac{\nabla\theta'}{r} - \frac{p_0}{\rho_0}\nabla\delta + \nu\nabla^2\mathbf{V}. \qquad (10.D.13)$$

The deviations from the equilibrium temperature θ' is governed by the equation

$$\nabla^2\theta' = 0, \qquad (10.D.14)$$

and hence can be expanded in a series (again neglecting dependence on the longitudinal angle) as

$$\theta' = \sum_l \left[A_l r^l + \frac{A'_l}{r^{l+1}}\right] p_l(\cos\theta). \qquad (10.D.15)$$

The lowest-order term in the series which gives a reasonable approximation to the actual differences in temperature as a function of latitude is the term $l = 2$, or

$$\theta' = \left(Ar^2 + \frac{A'}{r^3}\right)(1 - 3\cos^2\theta)$$
$$= T(r)(1 - 3\cos^2\theta) \qquad (10.D.16)$$

(recall that the polar angle θ is measured from the pole, while the latitude angle is measured from the equator—see Fig. 10.8). This equation, together with the Navier–Stokes equation and the equation of continuity, then determines the motion of the atmosphere.

If we pick the usual polar coordinates as shown in Fig. 10.8 and let

$$\beta = \frac{GM\rho_0\alpha}{a}$$

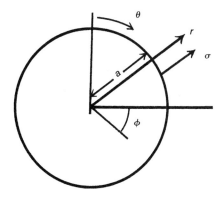

Fig. 10.8.   Polar coodinates for atmospheric circulation.

and

$$\frac{p_0}{\rho_0} = c^2,$$

the components of the Navier–Stokes equation are

$$c^2 \frac{\partial \delta}{\partial r} = \beta \frac{\partial \theta'}{\partial r} + \nu (\nabla^2 \mathbf{v})_r$$

and

$$\frac{1}{r} c^2 \frac{\partial \delta}{\partial \theta} = \frac{\beta}{r} \frac{\partial \theta'}{\partial \theta} + \nu (\nabla^2 \mathbf{v})_\theta, \tag{10.D.17}$$

while the equation of continuity is

$$\frac{1}{r^2} \frac{\partial}{\partial r} (r^2 v_r) + \frac{1}{r \sin \theta} \frac{\partial}{\partial \theta} (\sin \theta v_\theta) = 0. \tag{10.D.18}$$

Let us follow our usual line of attack and guess at a solution for these equations. From the form of θ', and the manner in which it appears in these equations, a reasonable guess might be

$$v_r = (1 - 3 \cos^2 \theta) f(r),$$
$$v_\theta = 6 \cos \theta \sin \theta \phi(r), \tag{10.D.19}$$
$$v_\phi = 0,$$

where $f(r)$ and $\phi(r)$ are to be determined.

The actual working out of the form of the functions $f(r)$ and $\phi(r)$ is straightforward, but tedious, and is left to Problem 10.3. We simply note that if we write

$$r = a + \sigma$$

and assume that

$$\frac{\sigma}{a} \ll 1$$

(this corresponds to taking the thickness of the atmosphere to be small compared to the radius of the earth), we find that

$$f(\sigma) = c \frac{\sigma}{8} (h - \sigma)(3h\sigma - \sigma^3),$$
$$\tag{10.D.20}$$
$$\phi(\sigma) = c \frac{\sigma}{48} (6h^2 - 15h\sigma + 8\sigma^2),$$

where

$$c = \frac{\alpha G M \rho_0}{\nu} \left(Aa^2 + \frac{A'}{a^3} \right). \tag{10.D.21}$$

and h is the height of the atmosphere.

It should be noted that the constants A and A' can easily be expressed in terms of the temperature difference between the pole and the equator by noting that the latter is

$$T_p = -2\left(Aa^2 + \frac{A'}{a^3}\right),$$

while the former is

$$T_e = Aa^2 + \frac{A'}{a^3},$$

so that

$$\frac{T_e - T_p}{3} = Aa^2 + \frac{A'}{a^3}. \tag{10.D.22}$$

What sort of winds does this solution describe? We note several things. First, we note that $\phi(\sigma)$ changes sign as σ goes from 0 to h. This means that as we go up in the atmosphere, v_θ must change sign. Since the factor $\cos\theta \sin\theta$ is always positive in the first quadrant, this means that for small σ, v_θ is positive (i.e. directed toward the equator), and for higher altitudes it is negative (i.e. directed toward the pole).

The function $f(\sigma)$ is positive definite as σ goes from 0 to h, which means that the radial velocity does not change as a function of altitude. However, at a critical value of $\theta_c (\approx 55°)$, the function $1-3\cos^2\theta$ changes sign. This means that from the equator to about 35° north latitude, v_r is positive, and the air is rising, while from this latitude to the pole, v_r is negative and the air is falling.

In the absence of rotation, then, the overall picture of the circulation which we have derived is shown in Fig. 10.9 and corresponds to the general picture which Hadley suggested two centuries ago.

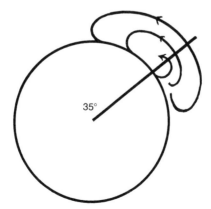

Fig. 10.9. The circulation corresponding to Eq. (10.B.20).

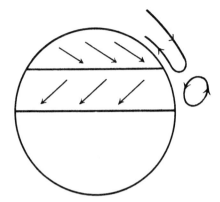

Fig. 10.10. The circulation which would result from the inclusion of the Coriolis force.

Had we included the Coriolis force in the Euler equation, the equations could be solved in the same way, although they are, of course, much more complicated. The general features of the solution are shown in Fig. 10.10. It turns out that the inclusion of rotation brings in a velocity in the ϕ-direction which greatly exceeds v_θ and v_r. This means that there will be easterly and westerly winds, rather than northerly and southerly. In fact, the trade winds and the prevailing westerlies follow this model, although it fails to predict the winds near the pole. To do better, it would be necessary to include higher-order terms in the temperature distribution in Eq. (10.D.15).

The model of the general circulation of the atmosphere bears less relation to the actual atmosphere than the simple theory of the tides presented in Chapter 6 does to actual tides. We have ignored many important effects besides rotation. These include the actual stratified structure of the atmosphere, the important effects of the presence of water vapor, and the effects of day–night temperature differences. Nevertheless, the reader should come away with the realization that many of the general features of atmospheric circulation can be understood in terms of the simple physical principles which we have introduced in this chapter.

SUMMARY

When a fluid is heated in the presence of a gravitational field, it is possible for an instability to occur, in which the warm fluid will rise and the cold fluid fall. This is called thermal convection. It will occur whenever an adverse temperature gradient exists in a nonviscous fluid, and whenever the Rayleigh criterion is met in a viscous one.

In the steady state, this type of motion, involving fluid rising at one point and falling at neighboring points, gives rise to the phenomenon of Bénard cells. The general circulation of the atmosphere can be thought of as being due to the unequal heating of the earth at the poles and equator. A simple model of the atmosphere was discussed.

PROBLEMS

10.1. Find the stability condition for a fluid of viscosity η between two free surfaces a distance h apart, with a temperature gradient $\beta = (\theta_2 - \theta_1)/h$,

in the case that the lower surface is solid, but the upper surface is free.

10.2. One of the current ideas about the structure of the earth is the theory of *continental drift*. The major idea of this theory is that the continental land masses are drifting around on top of convection cells in the mantle of the earth. Referring to Chapter 13 for typical sizes of the mantle, show that from what we have learned about Bénard cells that there should be roughly as many convection cells in the earth as there are continents observed.

10.3. Consider the assumed form of the solution to the equations of motion without rotation given in Eq. (10.D.17).

(a) Show that the equation of continuity in terms of the new function $f(r)$ and $\phi(r)$ is just

$$\frac{1}{r^2}\frac{\partial}{\partial r}(r^2 f(r)) - \frac{\phi(r)}{r} = 0.$$

(b) Eliminate δ from the two Euler equations to get a second equation relating $f(r)$ and $\phi(r)$.

(c) Assuming that Eq. (10.D.19) is valid, and applying the standard boundary conditions: namely

$$v_r = 0 \qquad \text{at } \sigma = 0, h$$

verify that Eq. (10.D.20) does indeed give the required solution to our problem.

10.4. For the case of an atmosphere with no rotation, calculate the variation of the atmospheric pressure with height at various latitudes. Make rough sketches of this variation.

10.5. Show that if a material contains heat sources (or sinks) which supply a quantity of heat Q per unit time, the heat equation (10.A.6) must be replaced by

$$\frac{D\theta}{Dt} = \kappa \nabla^2 \theta + \frac{Q}{\rho c_v}.$$

10.6. Consider a sphere of coefficient of diffusivity K immersed in a fluid of diffusivity K_2. If the sphere is of radius a, and a constant temperature gradient is maintained in the fluid, determine the temperature everywhere in the fluid and in the sphere.

10.7. Show that a law of similarity similar to that discussed in Chapter 9 for viscous flow can be derived for the steady-state flow of a viscous fluid of diffusivity K. To do this:

(a) Write down the equations of motion for the fluid.

(b) Show that if we define a new dimensionless number,

$$P = \frac{\nu}{\kappa}$$

called the Prandtl number, the temperature distributions in the fluid can depend on both R and P, while the velocity distribution can depend only on R (both, of course, can be functions of position).

(c) Hence, show that two flows are similar if the Reynolds and Prandtl numbers are equal.

10.8. In a similar way, show that for the type of convective processes which are discussed in the text, we can define a *Grashof number*

$$G = \frac{\alpha g L^3 (\theta_1 - \theta_0)}{\nu^2},$$

and that two convective flows will be similar if their Prandtl and Grashof numbers are equal. Why doesn't the Reynolds number enter into such considerations?

10.9. (a) Show that for small values of G, the heat transfer in a fluid must take place primarily through conduction, while for large values it must take place primarily through convection.

(b) Write the Rayleigh criterion [Eq. (10.B.28)] in terms of dimensionless numbers. For what relation between G and P will it be possible to have convection?

10.10. Consider the problem of the steady-state flow of a fluid which is confined to a vertical tube of radius R, with the upper end of the tube maintained at a temperature θ_2 and the lower end at a temperature θ_1.

(a) Assuming a form of the perturbation in which the velocity is along the z-axis (taken to be the axis of the tube), and v_z, θ' and dP'/dz depend only on the coordinates r and ϕ (the angle in the x-y plane), show that

$$\frac{1}{r}\frac{\partial}{\partial r}\left(r\frac{\partial}{\partial r}v_z\right) + \frac{1}{r^2}\frac{\partial^2 v_z}{\partial \phi^2} = G\frac{P v_z}{R^4}.$$

(b) Hence show that the velocities and temperature differences in this problem must be

$$v_z = v_0 \cos \phi [J_1(Kr)I_1(KR) - I_1(Kr)J_1(KR)],$$

$$\theta' = v_0 \left(\frac{\nu K^2}{\alpha g}\right) \cos \phi [J_1(Kr)I_1(KR) + I_1(Kr)J_1(KR)].$$

where $K = (GP/R^4)^{1/4}$.

10.11. For the geometry of Problem 10.10, find the stability criterion corresponding to Eq. (10.B.28).

10.12. In going from perturbation methods to steady-state solutions in Section 10.C, we simply set $n = 0$. Discuss the validity of this step in terms of your physical understanding of the meaning of the Reynolds number.

10.13. With the introduction of heat, we have still another form of energy which must be included in the type of energy balance carried out in Section 1.E. Show that for a viscous conducting fluid, conservation of energy requires that

$$\frac{\partial}{\partial t}\left(\frac{1}{2}\rho v^2 + \rho U\right) = -\frac{\partial}{\partial x_i}\left[\rho v_i\left(\frac{1}{2}v^2 + U + \frac{P}{\rho}\right)\right.$$
$$\left. - v_j\sigma'_{jk} - \kappa\frac{\partial\theta}{\partial x_i}\right],$$

where U is the sum of the internal energy defined in Eq. (10.A.3) and the usual potential energy.

10.14. Hence show that for the fluid of Problem 10.13,

$$\rho\theta\left(\frac{\partial S}{\partial t} + (\mathbf{v}\cdot\nabla)S\right) = \sigma_{ik}\frac{\partial v_i}{\partial x_k} + \frac{\partial}{\partial x_j}\left(\kappa\frac{\partial}{\partial x_j}\theta\right),$$

where S is the entropy density. Show that for an ideal fluid, this equation implies entropy conservation. It is called the *general equation of heat transfer*.

10.15. Consider a system which is made up of a mixture of two types of fluid, a normal fluid of density ρ_n and velocity $\mathbf{v}_n$ which is viscous and can carry entropy, and a superfluid of density ρ_s and velocity $\mathbf{v}_s$ which is nonviscous and carries no entropy.

(a) Show that the conservation of mass and entropy in such a fluid require

$$\frac{\partial\rho}{\partial t} + \nabla\cdot(\rho_n\mathbf{v}_n + \rho_s\mathbf{v}_s) = 0,$$

$$\frac{\partial s}{\partial t} + \nabla\cdot(s\mathbf{v}_n) = 0.$$

(b) Show that the Euler equation in such a fluid is

$$\frac{\partial}{\partial t}(\rho_n\mathbf{v}_n + \rho_s\mathbf{v}_s) + \frac{\partial}{\partial x_i}[(\mathbf{v}_s)_i\rho_s\mathbf{v}_s + (\mathbf{v}_n)_i\rho_n\mathbf{v}_n] = -\nabla P.$$

(c) Show that energy conservation in such a fluid requires

$$(\mathbf{v}_n - \mathbf{v}_s) \cdot \left[\frac{\partial}{\partial t} (\rho(\mathbf{v}_n - \mathbf{v}_s) + s\,\mathbf{v}\theta) \right] = 0.$$

10.16. For the fluid of Problem 10.15, show that if we treat v_n and v_s as small perturbations, and treat the derivatives of densities and thermodynamic quantities in the same way, we get

$$\frac{\partial^2 \rho}{\partial t^2} = \nabla^2 P$$

and

$$\frac{\partial^2 \rho}{\partial t^2} - \frac{\rho}{s} \frac{\partial^2 s}{\partial t^2} + \frac{\rho_s}{\rho_n} s \nabla^2 \theta = 0.$$

10.17. For the fluid of Problem 10.15, show that using the thermodynamic identities

$$\frac{\partial^2 s}{\partial t^2} = \left(\frac{\partial s}{\partial t} \right)_\rho \frac{\partial^2 \theta}{\partial t^2} + \left(\frac{\partial s}{\partial \rho} \right)_\theta \frac{\partial^2 \rho}{\partial t^2}$$

and

$$\nabla^2 P = \left(\frac{\partial P}{\partial t} \right)_\rho \nabla^2 \theta + \left(\frac{\partial P}{\partial \rho} \right)_\theta \nabla^2 \rho$$

along with the results of Problem 10.16 yields

$$\nabla^2 \rho - \frac{1}{c_1^2} \frac{\partial^2 \rho}{\partial t^2} + \gamma_1 \nabla^2 \theta = 0,$$

$$\nabla^2 \theta - \frac{1}{c_2^2} \frac{\partial^2 T}{\partial t^2} + \gamma_2 \frac{\partial^2 \rho}{\partial t^2} - 0,$$

where

$$c_1 = \sqrt{\left(\frac{\partial P}{\partial \rho} \right)_\theta}, \qquad c_2 = \sqrt{\frac{s \rho_s}{\rho \rho_n \left(\frac{\partial s}{\partial \theta} \right)_\rho}},$$

$$\gamma_1 = \left(\frac{\partial \rho}{\partial \theta} \right)_\rho, \qquad \gamma_2 = \frac{\rho_n}{s \rho_s} \left[1 - \frac{\rho}{s} \left(\frac{\partial s}{\partial \rho} \right)_\theta \right].$$

These results imply that there are two waves in the superfluid-normal mixture—a density wave, like the sound waves discussed in Chapter 5, and a thermal, or entropy wave which we have not seen before. This is called *second sound*, and is an important property of the type of *superfluids* which we have been discussing here.

10.18. How will winds on Venus differ from those on earth?

REFERENCES

C. Eckart, *Hydrodynamics of Oceans and Atmospheres*, Pergamon Press, New York, 1960.

A well-organized and detailed study of the general motions of oceans and the atmosphere. Contains an excellent discussion of the equations of motion including rotation.

B. Saltzman, *Theory of Thermal Convection*, Dover Publications, New York, 1962.

A collection of the classic papers on thermal convection. The original paper of Lord Rayleigh is the best presentation of the basic theory that I have found in the literature. There is a section on the motion of atmospheres.

S. Chandrasekar (cited in Chapter 3).

This book contains an exhaustive study of the effects of stability of rotation, magnetic fields, and viscosity, and is highly recommended for anyone wishing to read further in the field of thermal convection.

11

General Properties of Solids—
Statics

A. BASIC EQUATIONS

Up to this point, we have considered only one type of classical material—fluids. Fluids are characterized by the fact that on the microscopic level, the atoms interact mainly by collisions. The only forces which are generated inside of a fluid mass are those having to do with the momentum transferred through these collisions. We customarily refer to such forces as pressures. If we wished to apply an external force to a particular element in the fluid, however, it is clear that, aside from possible viscous drag, there is no way to generate forces inside the fluid which would oppose the applied force. Consequently, the fluid element would be in motion for as long as the force were applied.

If we think about a solid, however, we know that this is not true. If I push on a table top, the material immediately under my hand does not move (except, perhaps, for some small initial deformation which we will consider later). This means that the solid, unlike the fluid, *is* capable of generating internal forces which can oppose forces applied from the outside. The reason for this becomes clear if we think about the crystalline structure of the atoms in most solids. The atoms are locked into their places in a crystal lattice by electromagnetic interactions with other atoms, so that in order to move one atom, it is necessary to overcome the strong forces which bind it to other atoms (which, in turn, are bound to other atoms, and so forth). It is these atomic forces which we describe classically as "internally generated forces" in discussing solids, and which are absent in the case of fluids.

In our development of fluid mechanics, we found it simplest to discuss hydrostatics before hydrodynamics. We shall follow the same line here, and restrict our attention to the simple case in which a solid finds itself in static equilibrium with externally applied forces, leaving the problem of time-dependent effects for later. If equilibrium is indeed established, then the internally generated forces in the solid must exactly cancel the externally applied forces. Let us see how this idea leads us to the basic equations which describe the behavior of static solids.

Consider a solid (see Fig. 11.1) in which an external load $q(x)$ per unit length is applied externally. This external force might be the weight of the solid itself, or any combination of forces generated by the physical system. If we consider one infinitesimal volume element somewhere in the solid, then the forces acting on it in the y-direction are

(i) the loading, $q(x) \, dx$ acting at the center of the element,
(ii) an internally generated force $\tilde{F}$ acting on the left-hand edge, which we take to be acting in the positive direction (this force is written

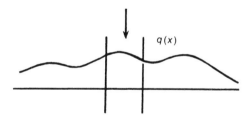

Fig. 11.1(a).   Loading of a solid.

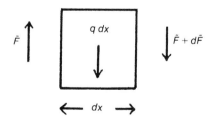

Fig. 11.1(b).   Vertical forces on an element in a loaded solid.

as $\tilde{F}$ in order to emphasize that it is an internally generated force, and *not* the force applied to the solid by an outside agency),

(iii) similar internally generated force acting on the right-hand edge, which we shall call $\tilde{F} + d\tilde{F}$, and assume acts in the negative y-direction.

Then balancing forces in the y-direction gives

$$\tilde{F} - q\,dx - (\tilde{F} + d\tilde{F}) = 0,$$

so that the rate of change of the internally generated force is related to the externally applied load in a solid by the equation

$$q(x) = -\frac{\partial \tilde{F}(x)}{\partial x}. \tag{11.A.1}$$

In addition to a balancing of forces in the y-direction, the requirement that the solid be in static equilibrium also demands that there be no net torque on the volume element. It follows from our discussion of the properties of a solid that a solid will be capable of generating internal torques, as well as internal forces. This, of course, is another difference between a solid and a liquid.

We can understand how torques might be generated by asking what happens to the loaded solid we considered above when a load is applied. Clearly, the solid will bend under the weight and deform, so that a volume element which started out as a cube, for example, would end up deformed as well (see Fig. 11.2). This deformation of the volume element must be

$q = 0$

$q = q(x)$

Fig. 11.2(a). The deformation of a loaded solid.

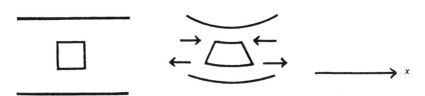

x

Fig. 11.2(b). Horizontal forces on an element in a loaded solid.

accomplished through the generation of forces in the x-direction such as those pictured, which act to compress the top of the cube and stretch the bottom. If we consider the net effect of the forces acting on one side of the volume element, it is to apply a torque to the element, attempting to cause the element to rotate. Thus, the internal forces generated in a direction perpendicular to the load can be represented by internally generated torques.

If we now consider a volume element, we see that there are three kinds of torques (see Fig. 11.3):

(i) those generated by internal forces in the x-direction,
(ii) those generated by internal forces in the y-direction,
(iii) those generated by the external load $q(x)$.

Balancing these torques about the point P leads to the result

$$\tilde{T} + (\tilde{F} + d\tilde{F})\,dx + q\cdot\frac{dx}{2}\cdot dx = \tilde{T} + d\tilde{T},$$

which becomes, when we drop terms of second order in infinitesimals,

$$\tilde{F} = \frac{\partial\tilde{T}}{\partial x}, \tag{11.A.2}$$

or, using Eq. (11.A.1)

$$q(x) = -\frac{\partial^2\tilde{T}}{\partial x^2}. \tag{11.A.3}$$

Once again, we write the internally generated torques as $\tilde{T}$ to distinguish them from externally applied torques.

These equations, which relate the internally generated forces to the externally applied load for a solid in static equilibrium, play the role of the

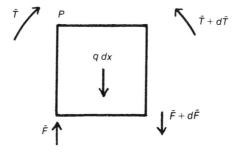

Fig. 11.3. Torques on a volume element in a deformed solid.

"equations of motion" for static solids. The internal forces, however, are not the type of things which one usually tries to calculate or measure. In Chapter 2, we were not interested in the internal forces operating in a star, but in the final shape of the star. Similarly, it is much more usual to ask how a given solid will deform when a load is applied than to ask about internal forces in the solid. Therefore, it is necessary to find some relation between the internal forces which we have calculated above the deformation in the solid.

B. HOOKE'S LAW AND THE ELASTIC CONSTANTS

The question of how much and in what manner a solid will deform under an applied force is an experimental one. There is no reason to expect, *a priori*, any particular kind of behavior. For example, if we imagined that the internal forces between the atoms in a solid could be represented by springs, then we might expect that the deformation would be proportional to the force applied. Such a solid is called an *elastic solid*, and will occupy most of our attention. We could also imagine that the forces between the atoms were such that they allowed no motion of the atoms unless the external force were strong enough to overcome them. In this case, there would be no deformation for small forces, and large deformations for large forces when, presumably, the material would fracture. Between these two extremes, one could imagine many different kinds of solids, and, indeed, there is an entire field of study called *rheology*, which is devoted to the study of the way in which materials react to forces applied to them.

For our purposes, however, we shall consider only the simple case of an elastic solid. To fix in our mind exactly what is meant by such a solid, imagine a thin wire of length l from which weights can be hung. For a given weight W, the wire will stretch a distance Δl. It is an experimental fact that for most materials, the amount of stretching is proportional to the force, so that

$$E \frac{\Delta l}{l} = W. \qquad (11.\text{B}.1)$$

This experimental finding is called *Hooke's law*, and the constant of proportionality E is called *Young's modulus*.

There is an interesting analogy between this law and a result which we found true for fluids in Chapter 1. The reader will recall that in order to specify the physical situation involving a fluid, it was necessary to say

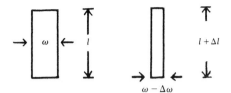

Fig. 11.4.   The deformation of a wire under a stress.

what kind of fluid we were considering—i.e. to specify an equation of state. Similarly, in this case, it is necessary to specify the type of solid being considered. To do this, it is necessary to give a relationship between the force applied to a solid and the amount of deformation suffered. Equation (11.B.1) is such a relationship, and hence plays the same role as the equation of state. Each elastic solid will be characterized by a different constant E, of course, just as different types of incompressible fluids are characterized by different densities.

It is also clear that if a solid is stretched in length, the material which goes to make up the extra length must come from somewhere. In general, if a solid is stretched in one dimension, it will thin down in the other dimension (see Fig. 11.4).

The ratio of the decrease in lateral dimension to the increase in length is called *Poisson's ratio*, and is defined by

$$\sigma = \frac{\Delta w}{\Delta l}. \tag{11.B.2}$$

The quantities E and σ are called *elastic constants*, because the two taken together completely specify the behavior of an elastic solid. In the next chapter, we shall discuss other sets of elastic constants (which can be related to E and σ) which are sometimes used for the same purpose. For the present, however, we shall work only with these two, and shall consider that we have completely specified the solid with which we are dealing if we have these two numbers.

C. BENDING OF BEAMS AND SHEETS

As a first example of the application of the laws derived in the preceding two sections, consider a thin beam of elastic material of Young's modulus E, originally straight, but bent by external forces into an arc of radius ρ (see Fig. 11.5).

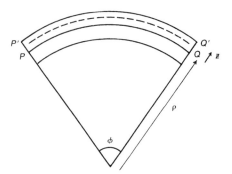

Fig. 11.5.   The bending of a filament.

In the process of bending, filaments near the top of the beam, such as $P'Q'$, will be stretched beyond their normal length, while those near the bottom will be compressed. There will be one filament, denoted by PQ, which is neither stretched nor compressed in the bending, but retains its normal length. This is called the neutral filament, and we will take ρ, the radius of curvature, to be measured from the center of curvature to the line PQ. The stretched length of $P'Q'$ is just

$$P'Q' = (\rho + z)\phi,$$

so that the fractional change in length of the filament $P'Q'$ is

$$\frac{P'Q' - PQ}{PQ} = \frac{(\rho + z)\phi - \rho\phi}{\rho\phi} = \frac{z}{\rho}. \tag{11.C.1}$$

But from the previous section, we know that this means that the force exerted on the filament $P'Q'$ must just be (by Hooke's law)

$$F = E\Delta l = E\frac{z}{\rho}. \tag{11.C.2}$$

Suppose we now look at the beam end on. The end of the filament $P'Q'$ will be an infinitesimal area element a distance z above the plane made up of the end points of the neutral filaments (see Fig. 11.6). Therefore, the torque being applied at this particular point is $Fz\,dA$, so that the total torque being applied to the end of the beam is

$$T = \int_A Fz\,dA = \int \frac{z^2 E}{\rho}\,dA = \frac{E}{\rho}I, \tag{11.C.3}$$

where $I = \int y^2\,dA$ is the moment of inertia of the cross section of the beam, and will depend on the shape of the cross section.

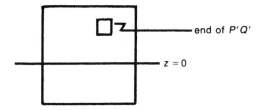

Fig. 11.6.   End view of a deformed filament.

We are now ready to relate the applied torque to the bending of the beam. Suppose that when there is no torque present, the beam is straight, and lies along the line $y = 0$. When the torque is applied, the beam will be deformed, and will be described by some curve $y(x)$ (see Fig. 11.7). From elementary calculus, the radius of curvature at each point along the beam will then be

$$\frac{1}{\rho} = \frac{d^2y/dx^2}{\sqrt{1 + (dy/dx)^2}} \approx \frac{d^2y}{dx^2}, \tag{11.C.4}$$

where the second approximate equality holds when the deformation of the beam is small (this is the only case which we shall consider). From Eq. (11.C.3), the applied torque at each point must then be

$$T = EI\frac{d^2y}{dx^2}. \tag{11.C.5}$$

In order for the beam to be in static equilibrium, this externally applied torque T must be canceled by the internally generated torque $\tilde{T}$ (i.e. we must have $T = -\tilde{T}$), so that the external loading of a beam (which is presumably what causes the torque in the first place) is related to the deformation by Eq. (11.A.2),

$$q(x) = \frac{d^2T}{dx^2} = EI\frac{d^4y}{dx^4}, \tag{11.C.6}$$

and, from Eq. (11.A.2), the external force is related to the deformation by

$$F = -EI\frac{d^3y}{dx^3}. \tag{11.C.7}$$

Fig. 11.7.   Side view of a deformed filament.

This is the general solution which we have been seeking. If we are told how much external force is applied to a solid (i.e. if we know $q(x)$), we can calculate the deformation at any point, $y(x)$ simply by solving Eq. (11.C.6). The solution of this equation will involve four integration constants, and these must be supplied by the boundary conditions. For example, if the problem were set up so that the end of the beam at $x = 0$ were free, then there would be no torques or forces at that end. This would give two conditions on the four constants. We will consider examples of other boundary conditions in subsequent sections, and in the problems.

Finally, we note that in all of our considerations so far, we have considered the beam to be made up of infinitely thin filaments which would stretch, but which had no lateral extent at all. This means that we have neglected the kind of effects which led up to the definition of Poisson's ratio in Eq. (11.B.2). A more realistic beam would both stretch and thin out as it was bent. In Problem 11.2, the reader will show that taking this effect into account leads to the equation for the deflection

$$q(x) = \frac{EI}{1 - \sigma^2} \frac{d^4 y}{dx^4}, \tag{11.C.8}$$

i.e. the result is equivalent to the one derived above except that the substitution $E \to E/(1 - \sigma^2)$ is made. Consequently, we will use Eq. (11.C.6) in all that follows, without losing any generality for the results.

D. THE FORMATION OF LACOLITHS

As an example of a physical situation in which the principles derived in the previous sections operate, consider the geological formation known as a lacolith. These occur when a fissure develops in a layer of rock below the surface of the earth, and molten magma under high pressure flows upward through this fissure, forcing the overlying layers of rock upward (see Fig. 11.8).

Fig. 11.8.   Schematic diagram of the formation of a lacolith.

There is a wealth of information on such formations, since they occur frequently. They are typically a mile or two across. There is one observed regularity to which we will turn our attention and that is the fact the higher the altitude of the lacolith, the smaller it will be. Let us see how the equations derived in the previous section can be applied to this problem.

We will consider the case where the fissure is a straight line which is very long compared to the width of the lacolith, so that we can ignore what happens at the ends. Then the forces on a strip of length b and width dx in the overburden (see Fig. 11.9) are

  (i) The weight of the rock pressing downward. If the density of the rock is γ, this will be $\gamma ab\,dx$.

 (ii) The force of the magma upward. This will be $Pb\,dx$, where P is the pressure of the fluid.

Thus, the net external force on the overburden per unit length in the x-direction is just

$$q(x) = b(P - \gamma a), \qquad (11.D.1)$$

so that the equation of deformation is

$$EI\frac{d^4 y}{dx^2} = b(P - \gamma a) \equiv \beta, \qquad (11.D.2)$$

which can be integrated to give

$$y = \frac{1}{24}\beta x^4 + \frac{1}{6}C_1 x^3 + \frac{1}{2}C_2 x^2 + C_3 x + C_4, \qquad (11.D.3)$$

where $C_1 \ldots, C_4$ are constants of integration.

As we pointed out in the previous section, these constants must be determined by the application of boundary conditions. At the point $x = 0$, we have $y = 0$ and $dy/dx = 0$ (this follows from the demand that there be

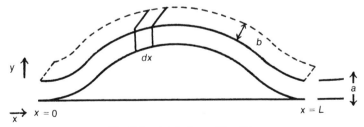

Fig. 11.9. A fully formed lacolith.

no discontinuity in the rock overburden). These conditions give

$$C_3 = C_4 = 0. \tag{11.D.4}$$

Similarly, and $x = L$, $y = 0$, and $dy/dx = 0$, so that

$$\frac{1}{24}\beta L^4 + \frac{1}{6}C_1 L^3 + \frac{1}{2}C_2 L^2 = 0,$$

$$\frac{1}{6}\beta L^3 + \frac{1}{2}C_1 L^2 + C_2 L = 0,$$

which gives

$$C_2 = \frac{\beta L^2}{12},$$

$$C_1 = -\frac{\beta L}{2}, \tag{11.D.5}$$

so that the equation describing the shape of the lacolith is

$$y = \frac{b(P - \gamma a)}{24EI}[x^4 - 2Lx^3 + x^2 L^2]. \tag{11.D.6}$$

This curve approximates those which are observed.

The height of the lacolith is given by the value of this function at its highest point, which from inspection is the point $x = L/2$. We find

$$h = \frac{b(P - \gamma a)L^4}{384EI}. \tag{11.D.7}$$

We can now explain the observed correlation of lacolith height with altitude which we cited earlier. Consider two lacoliths at different altitudes but fed from the same pool of magma (see Fig. 11.10). The difference in height ΔH between them will result in a difference in the

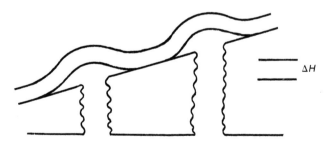

Fig. 11.10. Two lacoliths at different altitudes.

pressure of the magma. Since the height of the lacolith is directly proportional to the magma pressure, we would expect that the higher altitude lacolith would have the smaller height, as observed.

E. THE FORMATION OF MOUNTAIN CHAINS

Another geological phenomenon which we can understand on the basis of the physical principles presented in Section 11.C is the formation of mountain chains. In general, we can think of this process as a folding of the crust when a force is applied along the surface of the earth. This force might arise when the leading edge of a continent is pushed by continental drift against the underlying mantle. It is thought, for example, that the mountain chain on the west coast of North and South America was formed in this way. In general, a mountain chain will have the general shape shown in Fig. 11.11, where the largest mountains are closest to the applied force P, and the height of the mountains varies inversely with the distance from the force. There are, of course, exceptions to this general rule in nature, caused either by a nonuniformity in the crust or by deflections of the surface which exist before the force is applied.

As a model for this process, let us consider the crust to be a thin, semi-infinite sheet of material of Young's modulus E and moment of inertia I [see Eq. (11.C.6) and Problem 2.6] resting on top of an infinite elastic medium (see Fig. 11.12). The effect of this medium will be to exert

Fig. 11.11.   Deformation of a plate due to a horizontal force.

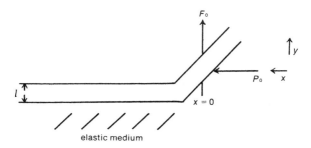

Fig. 11.12.   Model for the formation of a mountain chain.

a force which will oppose the deflection of the plate. For our purposes, we will treat this as a spring, so that the force exerted on any element of the sheet is proportional to the deflection of that element from equilibrium. It is clear that this is an effect which would be present in the cases of physical interest, where any attempt to push the crust down into the mantle would have to overcome the forces exerted by the mantle itself. We shall see that this type of restoring force, which we have not considered up to this point, is responsible for the inverse variation of mountain height with distance from the applied force.

Finally, we assume that some external force is applied at the end of the sheet. We label the force along the crust P_0 and the component of the force perpendicular to the crust by F_0. We include both of these forces because it is extremely unlikely that nature would ever provide a force *exactly* along the plane of the crust.

If the load per unit length on the crust is q, then the forces and torques acting on a section of the crust after equilibrium has been established are shown in Fig. 11.13.

In this diagram, P_0 is the applied external "axial" force, F the internally generated shear force, and $ky\,dx$ and $q\,dx$ the forces applied by the elastic medium and the loading, respectively. Balancing forces in the y-direction yields the equation

$$\tilde{F} - q\,dx - (\tilde{F} + d\tilde{F}) - ky\,dx = 0,$$

or

$$\frac{d\tilde{F}}{dx} = -q - ky, \tag{11.E.1}$$

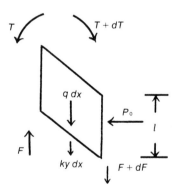

Fig. 11.13.   Forces and torques on a volume element in a mountain chain.

while balancing torques yields

$$-\tilde{T} + \frac{q}{2}(dx)^2 + \frac{ky}{2}(dx)^2 + \tilde{T} + d\tilde{T} - P\left(\frac{l}{2}\right) + P\left(\frac{l}{2} + dy\right) + (\tilde{F} + d\tilde{F})\,dx = 0,$$

or

$$\tilde{F} = -\frac{d\tilde{T}}{dx} - P\frac{dy}{dx}. \tag{11.E.2}$$

Using the previously derived relationship between deflection and torque [Eq. (11.C.3)], this becomes

$$EI\frac{d^4y}{dx^4} + P\frac{d^2y}{dx^2} + ky = q, \tag{11.E.3}$$

which is the working equation for our mountain chain model equivalent to Eq. (11.D.2) for the lacolith.

The general method of solving an inhomogeneous differential equation of this type is to note that to any particular solution, $y = y_p$ of Eq. (11.E.3) we can add y_h, a solution of the homogeneous equation

$$EI\frac{d^4y_h}{dx^4} + P\frac{d^2y_h}{dx^2} + ky_h = 0, \tag{11.E.4}$$

so that the most general solution is just

$$y = y_p + y_h. \tag{11.E.5}$$

[The reader can verify that this is indeed a solution of Eq. (11.E.3) by direct substitution.]

It is easy to see that the choice

$$y_p = \frac{q}{k} \tag{11.E.6}$$

satisfies Eq. (11.E.3). This is the particular solution discussed above, and is actually of little interest. It represents the amount the crust sinks into the mantle because of its own weight. We will ignore it in what follows.

The standard way to find y_h is to assume a solution of the form

$$y = e^{mx} \tag{11.E.7}$$

and determine the permissible values of m by direct substitution into Eq. (11.E.4). If we do this, and solve the resulting equation in m, we find

$$m^2 = \frac{-P \pm \sqrt{P^2 - 4EIk}}{2EI}, \tag{11.E.8}$$

which, if $P < 2\sqrt{EIk}$ (we will show in the next section that this is the only physically interesting case), gives complex values for m,

$$m = \alpha + i\beta,$$

where

$$\alpha = \left(\sqrt{\frac{k}{4EI}} + \frac{P}{4EI} \right)^{1/2} \tag{11.E.9}$$

and

$$\beta = \left(\sqrt{\frac{k}{4EI}} - \frac{P}{4EI} \right)^{1/2}.$$

The most general expression for y_h is thus

$$y_h = (C_1 e^{-\beta x} + C_2 e^{\beta x}) \cos \alpha x$$
$$+ (C_3 e^{-\beta x} + C_4 e^{\beta x}) \sin \alpha x. \tag{11.E.10}$$

As in the previous section, there are four constants which must be determined from the boundary conditions. Two constants can be determined directly from the requirement that the deflection be finite as x approaches $+\infty$ [x is measured positive to the left in the diagram above Eq. (11.E.1)]. This gives

$$C_2 = C_4 = 0.$$

Similarly, at $x = 0$, there is no external torque being applied, so that $T = -\tilde{T} = 0$, while the applied force F_0 must be equal and opposite to the internally generated force. Since

$$T = EI \frac{d^2 y}{dx^2},$$

the torque condition becomes

$$\left. \frac{d^2 y_h}{dx^2} \right]_{x=0} = 0,$$

which, substituting y_h from Eq. (11.E.10) leads to the result

$$C_1 = \frac{2\alpha\beta}{\beta^2 - \alpha^2} C_3. \tag{11.E.11}$$

In the same way, using the expression for the internally generated force in Eq. (11.E.2), together with Eq. (11.E.10) for y_h, gives

$$C_1 = \frac{-2F_0 \alpha\beta}{EI\alpha(3\beta^2 - \alpha^2)(\alpha^2 + \beta^2)}, \tag{11.E.12}$$

so that

$$y_h = \frac{F_0 e^{-\beta x} [-2\alpha\beta \cos \alpha x + (\alpha^2 - \beta^2) \sin \alpha x]}{\alpha \sqrt{kEI}(\alpha^2 + \beta^2)(3\beta^2 - \alpha^2)}. \tag{11.E.13}$$

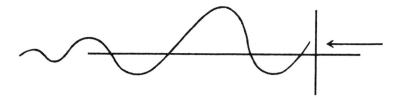

Fig. 11.14.   A typical shape for a mountain chain.

This is, of course, the equation of a damped oscillation, having a shape like that shown in Fig. 11.14. This is qualitatively the shape which we discussed for mountain trains in the beginning of this section. Once again, we see that by applying the simple ideas developed in the introductory sections of this chapter, the general features of a rather complicated system can be derived. Actually, the general equation (11.E.13) describes the distortion of any solid sheet which is subjected to an axial force and embedded in an elastic medium. Thus, in addition to describing a mountain train, it would also describe the folding of a vein of material embedded in other types of material—e.g., the bending of quartz veins embedded in harder rock.

Of course, there are many effects which we have ignored, so that Eq. (11.E.13) should be regarded as a first approximation to a correct description of a real mountain chain. In the problems, one such effect—the existence of initial deflections—is considered. Effects due to nonhomogeneity in the crust or mantle, fracturing of the rock or other nonelastic behavior, and nonuniform applied forces will not be discussed. There are, however, several special cases of Eq. (11.E.13) which are of considerable interest, and it is to these that we will turn in the next section.

F. SOME SPECIAL CASES: BUCKLING AND THE EULER THEORY OF STRUTS

There is a good deal of physics contained in Eq. (11.E.13). For example, it would seem from examining the equation if a force were applied to the solid sheet directly along the plane of the sheet (i.e. if F_0 were to vanish), there would be no deflection. Our intuition tells us in this case that there would be a force tending to compress the solid, but nothing to make it buckle. Our intuition also tells us that this would be a highly unstable situation, since the smallest force perpendicular to the sheet would produce a finite deflection. This is similar to the problem of the stability of stars which we considered in Chapter 3, where the smallest deviation from equilibrium could drive a system a long way when the equilibrium happened to be unstable.

To investigate the interesting aspects of this problem, let us begin by asking how large the axial load P can be for a given material. Imagine that the load P is applied in the presence of a small but finite F_0, and then gradually increased. What will happen?

Examining Eq. (11.E.13), we see that the deflection will be well behaved except when we approach the value of P which makes

$$3\beta^2 - \alpha^2 = 0. \tag{11.F.1}$$

For this value of P, the deflection will become infinite for any nonzero F_0. This phenomenon is known as "buckling" of the material. From Eq. (11.E.9), it will occur for a load P which satisfies

$$0 = 3\left(\sqrt{\frac{k}{4EI}} - \frac{P}{4EI}\right) - \sqrt{\frac{k}{4EI}} - \frac{P}{4EI}.$$

so that the critical load at which buckling occurs is just

$$P_{\text{crit}} = P = \sqrt{kEI}. \tag{11.F.2}$$

For loads below this, the deflection will remain finite, but as P approaches P_{crit}, the deflections will become arbitrarily large. Thus, material will support loads less than P_{crit}, but not greater. This was the origin of the statement following Eq. (11.E.8) that only values of P less than $2\sqrt{EIk}$ were physically interesting—higher loads would lead to buckling.

Of course, if F_0 were zero and P were at its critical value, Eq. (11.E.13) would be of an indeterminate form (zero divided by zero). Again, such a configuration would be highly unstable, and need not concern us further.

However, the astute reader will already have remarked that the critical loading given by Eq. (11.F.2) could not apply to a stressed material which was unconfined, because for such a system, the spring constant k would be zero. Unconfined beams which are required to carry an axial load are called "struts," and the study of their properties is, of course, of immense practical usefulness in construction of buildings, bridges, and other structures.

The theory of struts, first developed by Euler, is a special case of the problem treated in the previous section, but we see from Eq. (11.E.9) that if $k = 0$, $\beta^2 = -\alpha^2$ so that dividing by $\alpha^2 + \beta^2$, as we had to do to derive Eq. (11.E.13) will no longer be valid. In fact, it is probably easier to derive the deflection of a strut by starting from Eq. (11.E.8) directly than by finding some suitable limit of Eq. (11.E.13).

If the spring constant is zero, then the equation for m^2 is just

$$m^2(EIm^2 + P) = 0, \qquad (11.\text{F}.3)$$

so that

$$m = \pm i\sqrt{\frac{P}{EI}} = +i\gamma, \qquad (11.\text{F}.4)$$

which means that y will be given by an *undamped* oscillation, rather than a damped one. For the sake of definiteness, let us consider a strut of length L loaded with an axial load P at both ends (see Fig. 11.15).

The most general solution for y will then be

$$y = A \cos \gamma x + B \sin \gamma x.$$

To determine A and B it will be necessary to define boundary conditions. The most usual application of the theory of struts is in the case where both ends are held fixed, and the strut is compressed. This is the case we will consider here, and the case of a strut with free ends will be left to the problems.

If we require that $y = 0$ at $x = 0$ and $x = L$, we have

$$A = 0$$

and

$$B \sin \gamma L = 0,$$

which means that either

$$B = 0$$

or

$$\gamma = \frac{n\pi}{L} = \sqrt{\frac{P}{EI}}.$$

In the first case, the deflection is identically zero which means that the beam will not bend at all. If P approaches one of the critical values given by

$$P = EI \left(\frac{n\pi}{L}\right)^2,$$

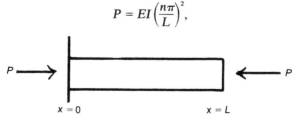

Fig. 11.15. A loaded strut.

however, then a solution of the form

$$y = B \sin\left(\frac{n\pi x}{L}\right) \tag{11.F.5}$$

is possible. Thus, the strut must either not bend at all, or be bent into an harmonic of a sine wave when the critical load is applied. In such a case, the value of B, the maximum deflection of the beam is undetermined, although in theory it is related to the force applied at the ends of the strut (see Problem 11.5).

This sudden transition to a deformed shape as P is increased is the analogue to the buckling of a member embedded in an elastic medium which we discussed above.

G. FENNO-SCANDIA REVISITED

In Chapter 8, we discussed the problem of viscous rebound in the context of the geological phenomenon of the Fenno-Scandian uplift—the rising of the crust of the earth after the melting of the glaciers. It was crucial to that discussion that the forces associated with the crust itself be negligible compared to the buoyant forces generated by the mantle under the crust. We are now in a position to show that this was a valid assumption.

For the sake of simplicity, consider an initial deformation of the crust given by (see Fig. 11.16)

$$y(t = 0) = y_0 \sin\left(\frac{\pi x}{L}\right) = \xi_0. \tag{11.G.1}$$

When the loading is lifted (e.g., when the glacier melts), the underlying fluid will exert a restoring force tending to lift the deformed part of the crust. The development of this process was treated in Section 8.C. For our purposes, we simply note that the restoring pressure at a point x is

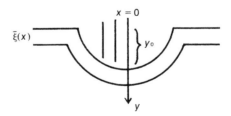

Fig. 11.16.   Elastic forces in the deformed crust.

simply the weight of the displaced fluid—i.e.

$$P = \rho g \xi, \tag{11.G.2}$$

where ξ is the y-coordinate of the surface, and at $t = 0$ is equal to ξ_0.

The question which we wish to answer concerns the relative importance of the force associated with this pressure and the force generated by the crust snapping back from its deformed position. We can attack this problem in the following way: Let us consider a section of crust which is initially flat, but loaded by a force

$$A\rho g y_0 \sin\left(\frac{\pi x}{L}\right) = q(x). \tag{11.G.3}$$

This force is proportional, with proportionality constant A, to the force exerted by the pressure of the mantle of the deformed crust.

If we consider the crust being deflected from a flat configuration by the applied force (11.G.3), then the amount of deformation for a given applied force will tell us how much force is required to produce a given deflection. In particular, we can ask what value of A is needed to produce a deflection equal to that produced by the glacier, and given by Eq. (11.G.1). If A turns out to be very small, we will have a situation in which forces very small compared to those of the actual pressure [Eq. (11.G.2)] will suffice to produce large deflections of the crust, while if A is large, it will take forces much greater than those associated with the pressure to deflect the crust. Since in the Fenno-Scandian uplift the crust is deflected only an amount ξ_0, our previous assumption that the crustal forces could be neglected would amount to an assumption that A be very small since in that case most of the buoyant forces must go into overcoming other things than forces generated in the crust.

From Eq. (11.C.6), we have

$$A\rho g y_0 \sin\left(\frac{\pi x}{L}\right) = EI\frac{d^4 y}{dx^4}, \tag{11.G.4}$$

so that, assuming that the crust is stationary at $x = 0$ and $x = L$, we find

$$y = A\left(\frac{L}{\pi}\right)^4 \frac{\rho g}{EI} y_0 \sin\frac{\pi x}{L}, \tag{11.G.5}$$

so that

$$\frac{y}{\xi_0} = A\left(\frac{L}{\pi}\right)^4 \frac{\rho g}{EI}. \tag{11.G.6}$$

If we now take parameters appropriate to Fenno-Scandia, namely $\rho = 3.27$ g/cc, $E = 10^9$ dynes cm^2, $L = 1400$ km, and evaluate I for a crust thickness of 35 km, we find (setting $y = \xi_0$)

$$A \approx \frac{1}{3600}. \tag{11.G.7}$$

This result means that in order to overcome forces generated within the crust by the initial deformation in Eq. (11.G.1), we would need a force which is only a small fraction of the actual buoyant force given in Eq. (11.G.2). Thus, virtually all of the force generated by the buoyancy must go into overcoming the viscous drag of the mantle (the process which we considered in Chapter 8) and almost none into overcoming the crustal forces themselves. This is what we set out to show.

SUMMARY

When a solid is subjected to external forces or torques, it generates within itself forces and torques which tend to oppose those being applied externally, and hence to bring the entire system into a state of static equilibrium. For the case of an elastic solid, a simple fourth-order differential equation can be written down which relates the amount of deformation of the solid to the magnitude of the external force.

Depending on the boundary conditions and the forces acting, this equation can be used to describe the general features of geological formations like lacoliths and mountain chains, or the buckling of struts when large axial loads are applied.

PROBLEMS

11.1. Consider a cantilever; i.e. a beam supported at one end only. Assume that the beam has Young's modulus E, moment of inertia I, weight per unit length q, and is of a length L.

(a) Write down the equation which describes the deformation of the beam as a function of length (see figure). This equation will have four undetermined coefficients.

(b) Write down the boundary conditions at $x = 0$, hence determine two of the four constants.

(c) Write down the boundary conditions at $x = L$, hence determine the remaining constants, showing that the deformation of the cantilever is given by

$$y = \frac{q}{EI}\left(\frac{x^4}{L^4} - \frac{Lx^3}{6}\right).$$

(d) Find the maximum deflection of the beam and the maximum *internal* torque generated in the beam.

11.2. Show that the equation equivalent to Eq. (11.C.6) for a solid beam is just

$$q(x) = \frac{EI}{1 - \sigma^2} \frac{d^4 y}{dx^4}.$$

11.3. In Section 11.D, we considered a lacolith formed by upward flow through a straight crack. Show that if we consider upward flow through a point hole in the lower strata, the equation which describes the shape of the surface layer is

$$y = \beta (R^2 - r^2)^2,$$

where R is the radius of the lacolith and r the radial distance from the center. What is the constant β?

11.4. Consider the formation of a mountain train in which an initial deformation is present. Take as a model a crust of moment I and Young's modulus E, embedded in a medium of spring constant k, and initially deformed to give a surface

$$y_0 = \delta_0 \sin \frac{\pi x}{L}.$$

(see figure).

Now suppose a load is applied to the ends of this deformation, as shown, so that the final configuration is

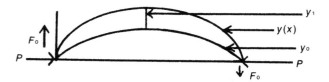

Assume all deformations are small, so that

$$y(x) = y_0 + y_1,$$

where y_1 is the extra bending due to the applied forces. Follow the steps leading to Eq. (11.E.13) to find the final shape of the crust. Will the phenomenon of buckling occur here?

11.5. Consider a strut with unsupported free ends, with an axial load P applied at $x = 0$ and $x = L$, and a force perpendicular to the strut at the ends be F_0 at $x = 0$, and $-F_0$ at $x = L$

(a) Show that the requirement that there be no torque at the ends leads to Eq. (11.F.4).

(b) Find a relation between P, F_0 and the undetermined constant B.

(c) Show that in the limit $F_0 = 0$, the only allowed solution for the strut is $y = 0$ everywhere. Interpret this result.

11.6. A solid beam of Young's modulus E and cross-sectional moment I is clamped at one end, and allowed to extend vertically in a gravitational field as shown. If the beam has weight $q(x)$ on it, and the internal forces act as shown:

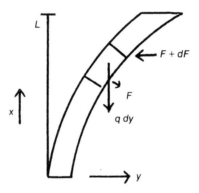

(a) Show that the equation for the deformation is

$$EI\frac{d^3y}{dx^3} = W(x)\frac{dy}{dx},$$

where

$$W(x) = \int^x q(x)\, dx.$$

(b) For the case $q(x) = q = $ const, show that this reduces to

$$\frac{d^2P}{dz^2} = -\beta zP,$$

where

$$\beta = \frac{q}{EI},$$

$$P = \frac{dy}{dx},$$

and

$$z = L - x.$$

(c) Show that a solution to this equation is

$$P = \alpha_0 + \sum_{n=1}^{\infty} \alpha_n z^n,$$

and find a relation between the α_n. From the fact that there is no torque at $z = 0$, show that we can write

$$P = \alpha_0 \left(1 - \frac{\beta z^3}{2 \cdot 3} + \frac{\beta^2 z^6}{2 \cdot 3 \cdot 5 \cdot 6} \cdots\right).$$

(d) At $z = L$, we must have $P = 0$. If $\alpha_0 \neq 0$, this means

$$0 = 1 - \frac{\beta L^3}{2 \cdot 3} + \frac{\beta^2 L^6}{2 \cdot 3 \cdot 5 \cdot 6}.$$

Show by plotting the right-hand side of the above as a function of βL^3 that there is a minimum value of βL^3 which will allow a solution ($\beta L^3_{min} \approx 7.8$). Hence find the maximum height to which a tree can grow. Do any trees in nature come close to this limit?

11.7. Show that the work done in stretching a length of filament dl as in Fig. 11.5 is just

$$\frac{1}{2} Ez^2 \, dA \frac{dl}{\rho^2}.$$

Hence, show that the energy stored in the bent beam is given by

$$W = \int_0^L \frac{T^2}{2EI} \, dl.$$

11.8. Hence show that the energy stored in the cantilever in Problem 11.1 is

$$W = \frac{q^2 L^5}{r0EI}.$$

Where did this energy come from?

11.9. Calculate the shape of a beam which is clamped at one end, and supported (but not clamped) at a level with the clamped end at its other end. Let the weight per unit length of the beam be q, and its length L.

11.10. Calculate the shape of a weightless beam clamped at both ends, but with a weight W applied at its center.

11.11. Consider a cantilever whose load per unit length is q, but which has a charge per unit length σ on it. Find the shape the cantilever will have in an electrical field E directed vertically. Will it ever curve up instead of down?

11.12. Carry through the analysis in Problem 11.8 when a constant force per unit length B is exerted in the horizontal direction on the beam. Hence discuss the effect of wind on vertical structures.

REFERENCES

As in the case of hydrodynamics, there are many standard texts on the theory of elasticity. These include

L. D. Landau and E. M. Lifschitz, *Theory of Elasticity*, Pergamon Press, New York, 1959.
   The same comments apply to this as to the Landau and Lifschitz text on hydrodynamics cited in Chapter 1.
A. E. H. Love, *A Treatise on the Mathematical Theory of Elasticity*, Dover Publications, New York, 1944.
   This text, like Lamb's *Hydrodynamics*, is an exhaustive treatment of many interesting and complicated problems, but suffers from a somewhat dated notational scheme.
John Prescott, *Applied Elasticity*, Dover Publications, New York, 1961.
   A book which has many worked examples of complicated systems without the advanced mathematics used in many texts.
Gerard Nadeau, *Introduction to Elasticity*, Holt, Rinehart, and Winston, New York, 1964.
   Uses a somewhat cumbersome dyadic notation, but discusses many simple problems in an understandable way.
I. S. Sokolnikoff, *Mathematical Theory of Elasticity*, McGraw-Hill, New York, 1956.
   Math is easy to follow, but there is little relation to experiment.
A. M. Johnson, *Physical Processes in Geology*, Freeman, Cooper, San Francisco, 1970.
   An excellent and readable account of the geological processes by which various formations are created. This book is especially valuable for physicists because of the clear treatment of descriptive geology which accompanies each example.
Carl W. Condit, *Scientific American*, Vol. 230 #2, p. 92, 1974.
   This is a very interesting discussion of wind bracing in tall buildings (see Problem 11.12).

See the references in Chapter 8 for readings on the Fenno-Scandian uplift.

12

General Properties of Solids— Dynamics

In England, we always let an institution strain until it breaks.

GEORGE BERNARD SHAW
Getting Married

A. THE STRAIN TENSOR

We have seen that the main feature which distinguishes solids from liquids is the ability to generate internal forces to oppose external loads placed upon them. In the case of elastic solids, these internal forces are related to the deformation of the solid via Hooke's law. Up to this point, however, we have not considered how a solid moves when a force is applied, but only the final static deformation. In order to discuss the dynamics of the response of a solid to forces, it will first be necessary to find a more general way of describing both the applied forces and the deformation of the solid.

Let us consider two points in a solid separated by a distance $d\mathbf{x}$ (see Fig. 12.1). After a force is applied, let the separation be $d\mathbf{x}'$. We can define the change in relative position by a vector $d\mathbf{u}$ in the equation

$$d\mathbf{x}' = d\mathbf{x} + d\mathbf{u}, \tag{12.A.1}$$

which means that the square of the separation is just

$$dx'^2 = dx'_i \, dx'_i = (dx_i + du_i)(dx_i + du_i)$$

$$= dx_i \, dx_i + 2 \frac{\partial u_i}{\partial x_k} \, dx_k \, dx_i + \frac{\partial u_i}{\partial x_k} \frac{\partial u_i}{\partial x_l} \, dx_k \, dx_l,$$

Fig. 12.1.   The deformation of a solid.

where we have used the relationship

$$du_i = \frac{\partial u_i}{\partial x_k} dx_k,$$

If we shift some dummy indices, and rearrange the second term, this can be written as

$$dx'^2 = dx^2 + 2u_{ik} dx_i dx_k, \tag{12.A.2}$$

where the tensor u_{ik} is just given by

$$u_{ik} = \frac{1}{2}\left(\frac{\partial u_i}{\partial x_k} + \frac{\partial u_k}{\partial x_i} + \frac{\partial u_i}{\partial x_k}\frac{\partial u_i}{\partial x_k}\right). \tag{12.A.3}$$

This is an extremely important quantity, as we shall see, and is called the *strain tensor*.

From the derivation, it is clear that the strain tensor must describe the deformation of a solid. Since the derivatives of u_i describe the change of relative coordinates in the solid [see Eq. (12.A.2)], so long as we confine our attention to small deformations (as we did in Chapter 11), we can drop second-order terms in the derivatives of u_i, so that

$$u_{ik} = \frac{1}{2}\left(\frac{\partial u_i}{\partial x_k} + \frac{\partial u_k}{\partial x_i}\right). \tag{12.A.4}$$

This is the form which we shall use throughout the remainder of the discussion.

In order to understand what the strain tensor means, we shall look at two examples of strain tensors and deduce the actual deformations to which they correspond. For the first example, consider a strain tensor given by

$$u_{ik} = \begin{pmatrix} u_1 & & 0 \\ & u_2 & \\ 0 & & u_3 \end{pmatrix}.$$

In such a diagonal strain tensor, the values of dx' are given by

$$dx'_i = (1 + u_i) dx_i,$$

so that the volume of an element on the solid after the deformation would just be

$$dV' = dx'_1 \, dx'_2 \, dx'_3 = dx_1 \, dx_2 \, dx_3(1 + u_1 + u_2 + u_3)$$
$$= dV(1 + u_{ii}),$$

where we have, again, dropped second-order terms in the deformation. The quantity u_{ii} is the trace of the strain tensor, and we see that the relative change of volume of an element is given by

$$\frac{dV' - dV}{dV} = u_{ii}. \tag{12.A.5}$$

Thus, from our first example, we see that the diagonal elements of the strain tensors are related to changes in volumes in the solid, and that a purely diagonal strain tensor corresponds to either a compression or dilation of the solid.

For our second example, consider a strain tensor which has only off-diagonal elements, such as

$$u_{ik} = \begin{pmatrix} 0 & u_{12} & 0 \\ u_{21} & 0 & 0 \\ 0 & 0 & 0 \end{pmatrix}$$

(since the tensor is symmetric, $u_{12} = u_{21}$). Consider a solid which before deformation contains two vectors (see Fig. 12.2) A and B, initially along the 1 and 2 axes, respectively. After the deformation, the vectors will have shifted over, and we will have (by definition)

$$\delta A_2 = u_{21}A,$$
$$\delta B_1 = u_{12}B.$$

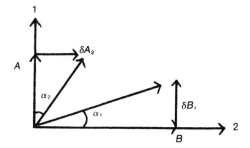

Fig. 12.2. Rotation of position vectors in a solid.

For small angles, the angles through which each vector has been rotated are given by

$$\alpha_2 \approx \frac{\delta A_1}{A} = u_{12},$$

$$\alpha_1 = \frac{\delta B_2}{B} = u_{21},$$

so that the total change in angle between the two vectors is just

$$\alpha = \alpha_1 + \alpha_2 = 2u_{12}. \tag{12.A.6}$$

Thus, we see that the off-diagonal elements of the strain tensor are related to shear deformations in the solid, and, in fact, can be related to the shear angle α.

The strain tensor, then, gives us a way of describing the most general kinds of deformations which can take place in a solid. The diagonal elements correspond to the compression or dilation of the solid, while the off-diagonal elements correspond to shearing.

B. THE STRESS TENSOR

Now that we have developed a generalized way of describing the deformation of a solid when forces are applied, we need to develop an equally generalized way of describing the forces themselves. This is known as the stress tensor, and has already been discussed (although not under this name) in Chapter 8, where the tensor σ_{ik} was introduced to describe the viscosity in a fluid [see Eq. (8.A.3)].

Let us introduce the idea of a stress tensor by noting that when a body is deformed, each infinitesimal element in the body feels a force per unit volume F exerted on it by its neighbor. For an element in the interior of the body, these forces will cancel out in the static case, but for an element in the surface, they will not (see Fig. 12.3). This, of course, is the mechanism by which a force is generated at the surface of a body to cancel the applied forces [see Eqs. (11.C.6) and (11.C.7)]. The reader may be interested in comparing this idea of force cancellation in the body of a solid with the idea of surface tension in a fluid (Chapter 5) or in a nucleus (Chapter 7).

In complete analogy to the development in Section 8.A, we can write

$$F_i = \frac{\partial \sigma_{ik}}{\partial x_k} \tag{12.B.1}$$

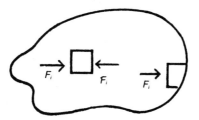

Fig. 12.3.   Forces on internal and surface volume elements in a solid.

for the force per unit volume in the ith-direction on a volume element, so that the total force in the ith-direction is just

$$F_i = \int_V \frac{\partial \sigma_{ik}}{\partial x_k} \, dV = \int_S \sigma_{ik} \, dS_k. \tag{12.B.2}$$

The tensor σ_{ik} whose divergence is the body force in a solid is called the *stress tensor*, and is extremely important in the discussion of solids. We see that it can be interpreted [Eq. (12.B.2)], as the net force in the ith-direction on a surface perpendicular to the kth-direction. To make this idea clear, consider Fig. 12.4 in which a surface in the y-z plane is drawn. This surface is perpendicular to the x-direction. In general, there are three types of forces that can be exerted on it

  (i) a force directed along the x-direction, which we would term σ_{xx} ;
  (ii) a force along the y-direction, called σ_{yx} ;
  (iii) a force along the z-direction, called σ_{zx}.

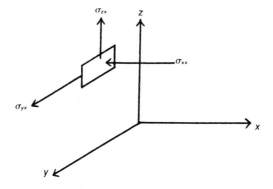

Fig. 12.4.   An interpretation of the stress tensor.

The first of these is a compressional force, while the other two are shear forces.

There are two points to note about the stress tensor before we look at some examples. First, we note again that in describing a force acting on a surface, two things must be specified: The direction of the force and the direction of the surface. This is why a second rank tensor provides the most natural description of forces acting at the surfaces of solids.

Second, in almost all of the problems which are encountered in dealing with solids, the stress tensor is given. Just as in dealing with static deformations we were given the load and had to discover the shape of the material, in the more general problems which we shall describe, the forces acting on a solid will be given, and we shall want to find the response of the solid (described by the strain tensor). This is exactly analogous to the usual problem in mechanics, in which we are given the forces acting on a body, and then required to find the subsequent motion. It is easy to show [see Problem(12.3)] that the stress tensor must be symmetric—i.e. that

$$\sigma_{ik} = \sigma_{ki}.$$

Further familiarization with σ_{ik} is probably best done through examples.

Consider first a body immersed in a fluid, which exerts a pressure P on the surface. Since the pressure by definition acts perpendicular to the surface, the force exerted on a surface element is simply

$$F_i = - P \, dS_i = - P\delta_{ik} \, dS_k,$$

where the minus sign denotes an inward force. For this case, the stress tensor is just

$$\sigma_{ik} = - P\delta_{ik},$$

so that a pure compression of dilation corresponds to a diagonal stress tensor.

Another important example of a stress tensor can be taken from the field of electricity and magnetism (the reader unfamiliar with this field can skip ahead to Section 12.C without loss of continuity). The electricity and magnetic fields, $\mathbf{E}$ and $\mathbf{B}$, are given in terms of charge and current densities, ρ and $\mathbf{j}$, by the Maxwell equations

$$\nabla \cdot \mathbf{E} = 4\pi\rho, \tag{12.B.3}$$

$$\nabla \cdot \mathbf{B} = 0, \tag{12.B.4}$$

$$\nabla \times \mathbf{B} = \frac{4\pi}{c}\mathbf{j} + \frac{1}{c}\frac{\partial \mathbf{E}}{\partial t}, \tag{12.B.5}$$

$$\frac{1}{c} \frac{\partial \mathbf{B}}{\partial t} = -\nabla \times \mathbf{E}. \tag{12.B.6}$$

Suppose that we had a collection of charges and currents enclosed in a volume V. Then the total force acting on the charges and current would just be

$$\mathbf{F} = \int_V \left(\rho \mathbf{E} + \frac{1}{c} \mathbf{j} \times \mathbf{B} \right) dV.$$

Using Eq. (12.B.3) to eliminate ρ and Eq. (12.B.5) to eliminate $\mathbf{j}$ from this expression, we have, after adding and subtracting

$$\frac{1}{c} \left(\mathbf{E} \times \frac{\partial \mathbf{B}}{\partial t} \right)$$

to the integrand

$$\mathbf{F} + \frac{1}{4\pi c} \frac{\partial}{\partial t} \int_V (\mathbf{E} \times \mathbf{B}) \, dV = \frac{1}{4\pi} \int \left[\mathbf{E}(\nabla \cdot \mathbf{E}) + \frac{1}{c} \left(\mathbf{E} \times \frac{\partial \mathbf{B}}{\partial t} \right) - \mathbf{B} \times (\nabla \times \mathbf{B}) \right] dV.$$
$$\tag{12.B.7}$$

The left-hand side of this equation now is a force (i.e. the time derivative of the momentum of the particles) and the time derivative of $(1/4\pi c)(\mathbf{E} \times \mathbf{B})$, which we identify as the momentum of the field. Thus, the left-hand side is just the time rate of change of the total momentum. The right-hand side, on the other hand, can be rewritten using the result of the first vector identity in Problem 1.1 and Eq. (12.B.4) to read

$$\frac{d\mathbf{P}_T}{dt} = \frac{1}{4\pi} \int_V [\mathbf{B}(\nabla \cdot \mathbf{B}) + (\mathbf{B} \cdot \nabla)\mathbf{B} - \tfrac{1}{2}\nabla(\mathbf{B} \cdot \mathbf{B}) + \mathbf{E} \cdot (\nabla \cdot \mathbf{E})$$
$$+ (\mathbf{E} \cdot \nabla)\mathbf{E} - \tfrac{1}{2}\nabla(\mathbf{E} \cdot \mathbf{E})] \, dV,$$

which, in Cartesian tensor notation, is just

$$\frac{\partial P_i}{\partial t} = \frac{1}{4\pi} \int \frac{\partial}{\partial x_j} [B_i B_j + E_i E_j - \tfrac{1}{2}\delta_{ij}(E^2 + B^2)] \, dV,$$

which is precisely the form of Eq. (12.B.2).

If we write

$$T_{ij} = \frac{1}{4\pi} [E_i E_j + B_i B_j - \tfrac{1}{2}\delta_{ij}(E^2 + B^2)], \tag{12.B.8}$$

then T_{ij} is the *Maxwell stress tensor*, a quantity familiar from electrodynamics. It is, of course, just one example of a stress tensor, and we have calculated it by determining directly the forces acting on each point of our body.

It is important to emphasize that in calculating the stress tensor, we are adding nothing to our knowledge of the physics of the system. We are simply rewriting the statements about the forces acting on a body in a way which shall turn out to be very convenient for us.

C. EQUATION OF MOTION FOR SOLIDS

Having now defined the stress and strain tensors, we have at our disposal completely general ways of describing both the forces which are applied to a solid and the way in which the solid deforms in response to these forces. The problem now is to relate these two descriptions—i.e. to find the deformation in a given solid corresponding to a given force.

There is no *a priori* relation between the stress and the strain. Such a relation depends entirely on the material being stressed. We can easily imagine many different kinds of response to an applied force. For example, we know that in some cases (see Section 2.B) we can talk about elastic solids, where the deformation is directly proportional to the stress. We might picture the microscopic structure of the material as in Fig. 12.5, where the atoms are held together by springs. When a force is applied, the material will deform until the springs have compressed enough to counteract the applied force F.

On the other hand, we could imagine a material in which the atoms were held together by rigid rods, so that if a force is applied, the solid does not deform at all until the applied force reaches the point where it can break the rods (see Fig. 12.6). We would then see the material fracture instantaneously.

A third possible kind of response to a force would be one in which the

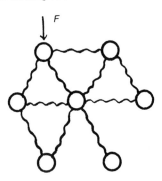

Fig. 12.5.   An harmonic solid.

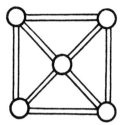

Fig. 12.6.   A rigid solid.

force is related to the *rate* of deformation (i.e. to the time derivatives of the strain tensor). We saw forces of this type in Chapter 7 when we discussed viscosity, which gave rise to a force which depended on the velocity of the fluid. We might picture such a solid as one in which the bonds between the atoms are very weak, so that a force which is applied continuously results in a continuous deformation. Such a system would be called a *Newtonian solid*, and is discussed in Problem 12.4.

The point of this exercise is to illustrate the remark made at the beginning of this section—there is no way we can tell from first principles how a solid will respond to an applied force. This is an exact logical analogy to the lesson we learned in Section 1.D, when we found that in dealing with a fluid system, we had to have an equation of state, which told us what sort of fluid we had in the system. In the case of a solid, specifying the type of material in the system corresponds to giving a relation between the stress and the strain.

Throughout the rest of the text, we shall be concerned primarily with elastic solids. In such solids, we expect that the stress will be proportional to the strain. Since σ_{ik} is a symmetric tensor, and since it must be proportional to the strain tensor, the most general form of the stress-strain relationship must be

$$\sigma_{ik} = \lambda u_{ll}\delta_{ik} + 2\mu u_{ik}, \tag{12.C.1}$$

where λ and μ are called the Lamé coefficients and differ from one elastic solid to another. Equation (12.C.1) is simply Hooke's law in tensor form, as will be verified explicitly later.

Thus, in order to supply the "equation of state" for a solid, we must

1. give a stress-strain relationship, which tells us what general class of solids we have [Eq. (12.C.1) defines a general elastic solid],
2. specify the coefficients which say which particular solid in that

general class we have (e.g., giving λ and μ completely specifies the elastic solid).

In order to make the tensor from Hooke's law a little more familiar, let us look at some examples. First, consider a cylinder under a tension T, so that the stress tensor is

$$\sigma_{11} = T,$$
$$\sigma_{ij} = 0 \quad \text{otherwise.} \tag{12.C.2}$$

Then the three diagonal equations from Eq. (12.C.1) are

$$\sigma_{11} = T = \delta_{11}\lambda u_{ll} + 2\mu u_{11},$$
$$\sigma_{22} = 0 = \delta_{22}\lambda u_{ll} + 2\mu u_{22}, \tag{12.C.3}$$
$$\sigma_{33} = 0 = \delta_{33}\lambda u_{ll} + 2\mu u_{33}.$$

Adding these three equations, and recalling that

$$u_{ll} = u_{11} + u_{22} + u_{33},$$

we have

$$u_{ll} = \frac{T}{3\lambda + 2\mu}, \tag{12.C.4}$$

which, if we plug back into Eqs. (12.C.3) yields

$$u_{11} = \frac{T(\lambda + \mu)}{\mu(3\lambda + 2\mu)} \tag{12.C.5}$$

and

$$u_{22} = u_{33} = -\frac{\lambda T}{2\mu(3\lambda + 2\mu)} \tag{12.C.6}$$

for the three diagonal elements of the strain tensor.

Now u_{11} is the deformation of the cylinder along the direction of the tension. By definition, this is related to the tension by

$$T = Eu_{11},$$

where E is Young's modulus. Similarly, Poisson's ratio is

$$\sigma = \frac{u_{22}}{u_{11}},$$

which, upon substitution, yields

$$\lambda = \frac{E\sigma}{(1 + \sigma)(1 - 2\sigma)} \tag{12.C.7}$$

and

$$\mu = \frac{E}{2(1+\sigma)}. \tag{12.C.8}$$

We see, then, that the Lamé coefficients are simply related to E and σ, the numbers which we used in Chapter 11 to define an elastic material.

There are another set of constants which are often used as alternatives to λ and μ or E and σ in describing elastic solids. To understand these, consider two examples: First, consider a pure shearing force, so that

$$\sigma_{12} = \sigma_{21} = T,$$
$$\sigma_{ij} = 0 \quad \text{otherwise.} \tag{12.C.9}$$

Then we have

$$\sigma_{12} = T = 2\mu u_{12},$$

so that, recalling Eq. (12.A.6)

$$2u_{12} = \alpha = \frac{T}{\mu},$$

which can be used to define the *shear modulus* (the proportionality constant between the applied shear and the angle of deformation) as

$$\mu = \frac{T}{\alpha}, \tag{12.C.10}$$

which is, of course, identical with the Lamé coefficient μ.

Next consider a solid under hydrostatic compression, so that

$$\sigma_{ij} = -P\delta_{ij}. \tag{12.C.11}$$

If we then follow the exact steps of Eq. (12.C.3) to Eq. (12.C.4), we find

$$u_{ll} = -\frac{3P}{3\lambda + 2\mu} = \frac{\Delta V}{V},$$

where we have used Eq. (12.A.5), a general property of the strain tensor. We can then see that the ratio of volume change to applied pressure is just

$$\frac{-\dfrac{\Delta V}{V}}{P} = \frac{1}{\lambda + \frac{2}{3}\mu} = K, \tag{12.C.12}$$

which is usually called the *bulk modulus* of the material.

In what follows, then, we shall feel free to use any of these three sets of elastic constants to define our solid, depending on which is most convenient in a particular problem.

Having now written down Hooke's law and seen what the Lamé coefficients represent in terms of deformations of a solid, we can turn to the problem of writing down the equation of motion for an infinitesimal element in the solid. This is analogous to deriving the Euler equation, since both involve Newton's second law. If f_i is the force in the ith-direction on an infinitesimal volume element, then Newton's second law for that volume element is

$$f_i = \rho \frac{\partial^2 u_i}{\partial t^2}. \tag{12.C.13}$$

But we know that [see Eq. (12.B.1)]

$$f_i = \frac{\partial \sigma_{ik}}{\partial x_k},$$

so that

$$\frac{\partial}{\partial x_k} [\lambda u_{ll}\delta_{ik} + 2\mu u_{ik}] = \rho \frac{\partial^2 u_i}{\partial t^2}. \tag{12.C.14}$$

If we use the definition of the strain tensor [Eq. (12.A.4)] and rearrange terms, this becomes

$$(\lambda + \mu) \frac{\partial}{\partial x_i} \left(\frac{\partial u_k}{\partial x_k}\right) + \mu \frac{\partial^2 u_i}{\partial x_k^2} = \rho \frac{\partial^2 u_i}{\partial t^2} \tag{12.C.15}$$

or, in vector form

$$(\lambda + \mu)\nabla(\nabla \cdot \mathbf{u}) + \mu \nabla^2\mathbf{u} = \rho \frac{\partial^2 \mathbf{u}}{\partial t^2}. \tag{12.C.16}$$

This is the basic equation which describes the time-dependent response of an element in an elastic solid to an applied force. The remainder of this chapter will be devoted to examining the consequences of the equation. Before moving on, however, we shall, for the sake of completeness, write down two forms of the equation which shall be useful later. If we take the gradient of Eq. (12.C.16), we find that

$$\frac{\partial^2}{\partial t^2} (\nabla \cdot \mathbf{u}) = \frac{\lambda + 2\mu}{\rho} \nabla^2(\nabla \cdot \mathbf{u}), \tag{12.C.17}$$

while if we take the curl of the equation, we have

$$\frac{\partial^2}{\partial t^2} (\nabla \times \mathbf{u}) = \frac{\mu}{\rho} \nabla^2(\nabla \times \mathbf{u}). \tag{12.C.18}$$

D. BODY WAVES IN ELASTIC MEDIA

In the case of fluids, we saw that a great deal of interesting information could be derived by looking for wave-type solutions of the equations of motion. It is interesting to ask whether the same is true for solids. We shall see that there are several different types of waves which can propagate through a solid, and we shall see how this information has enabled us to discover the composition of the interior of the earth through the development of the science of seismology

From Eqs. (12.C.17) and (12.C.18), it is clear that waves will exist. Rather than proceed formally from the equations of motion, however, let us look at examples of two different types of waves and simply verify that they satisfy the equations of motion for a solid.

We shall first look for solutions of the equation of the following type: A wave disturbance of some sort travels in the x-direction, and the displacement of the solid is in the x-direction as well. This corresponds to

$$u_x = u_x(x - ct),$$
$$u_y = u_z = 0, \qquad\qquad (12.D.1)$$

so that

$$\nabla \cdot \mathbf{u} = \frac{\partial u_x}{\partial x},$$

and the equation of motion becomes

$$\rho \frac{\partial^2 u_x}{\partial t^2} = (\lambda + \mu) \frac{\partial}{\partial x}\left(\frac{\partial}{\partial x} u_x\right) + \mu \frac{\partial^2 u_x}{\partial x^2},$$

which reduces to

$$\frac{\partial^2 u_x}{\partial t^2} = \frac{\lambda + 2\mu}{\rho} \frac{\partial^2 u_x}{\partial t^2}. \qquad\qquad (12.D.2)$$

This, of course, is the equation of a wave traveling along the x-axis with velocity c_l, where

$$c_l^2 = \frac{\lambda + 2\mu}{\rho}. \qquad\qquad (12.D.3)$$

We have used the subscript l because this is a longitudinal wave, since the displacement is in the same direction as the velocity of the wave. It can be interpreted as a compressional wave as well. To see this, let us plot u_x as a function of x for fixed t. In Fig. 12.7, u_x positive corresponds to

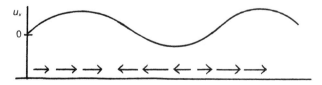

Fig. 12.7. Velocities of particles for acoustic waves.

the particles moving to the right (which we take to be the direction of the wave), and u_x negative corresponds to particles moving to the left. In the lower part of the figure we show the actual direction of motion of elements. We see that elements of the solid tend to move toward every other point where u_x is zero, and away from the points where u_x is an extremum. Thus, the density in the former regions will be greater than the density around the latter. This will be observed as a pattern of density variations which, as time progresses, will move to the right. This is just what a sound wave is, and hence this type of wave is sometimes called an acoustic wave.

A more usual type of wave is the transverse wave, in which the displacement of the material is perpendicular to the direction of motion of the wave. A wave on a string would be an example of such a phenomenon. This type of wave, should it exist, would correspond to

$$u_y = u_y(x - ct),$$
$$u_z = u_x = 0,$$

(12.D.4)

which gives

$$\nabla \cdot \mathbf{u} = 0,$$

so that the equation of motion in the y direction is just

$$\frac{\partial^2 u_y}{\partial t^2} = \frac{\mu}{\rho} \frac{\partial^2 u_y}{\partial x^2},$$

(12.D.5)

which is again a wave equation for a transverse wave (the terms longitudinal and transverse refer to the direction of the displacement relative to the direction of motion of the wave). The velocity of the wave is

$$c_t^2 = \frac{\mu}{\rho}.$$

(12.D.6)

This type of transverse wave in a solid is called a *shear wave*, since 't can be thought of as a small-scale shearing in the body of the solid.

It is easy to see that each of these waves corresponds to a different form of the equation of motion. The compressional wave corresponds to volume changes in the solid, and hence to Eq. (12.C.17), while the shear wave corresponds to Eq. (12.C.18). One important consequence of the fact that there are two types of waves which can be excited in a solid, each traveling with a different speed, is in seismology. To see this, we note that from Eqs. (12.D.3) and (12.D.6) that

$$c_l > c_t,$$

so that the compressional wave travels faster than the shear wave. If we imagine a disturbance somewhere deep in the earth, with both compressional and shear waves coming out, the compressional wave will reach the surface first. Hence, seismologists refer to it as the P wave, or principle wave. The shear wave arrives at some later time, and hence is called the S, or secondary wave. Therefore, there would be two shocks arriving at the surface after such a disturbance, and the time difference between their arrivals would depend on the relative values of c_l and c_t. These, in turn, depend on the density and the kind of material of which the earth is composed.

By measuring the time lag between the arrival of different waves from a disturbance, one can obtain information about the structure of the earth. This is the aim of the science of seismology, which we shall discuss later.

As an example of this effect, let us consider an earthquake at Tokyo, and ask what the time difference is between the P and S waves as observed at San Francisco. Let us assume the earth has a uniform density, and that the coefficients are everywhere constant and are equal to those for fused silicates. These waves will travel directly across a chord of the earth, which is 9.5×10^6 m long. Therefore, the time difference will be

$$\Delta t = l \sqrt{\rho} \left(\frac{1}{\sqrt{\lambda + 2\mu}} - \frac{1}{\sqrt{\mu}} \right) \approx 25 \text{ min.}$$

As we shall see later, the fact that the composition of the earth varies as a function of depth makes the actual calculation of the paths of seismic waves and of the properties of the earth's interior quite a bit more difficult.

E. SURFACE WAVES IN SOLIDS

In Chapter 5, we saw that it is possible to have waves in a fluid which exist only in the surface, and which die out rapidly as a function of depth. In this section, we shall see that such waves can exist in solids as well. Unlike the P and S waves which we considered in the last section, the existence of surface waves depends on applying both the equation of motion and the boundary conditions.

Consider a semi-infinite solid (as shown in Fig. 12.8) in which a wave propagates with velocity

$$c = \frac{\omega}{k}. \tag{12.E.1}$$

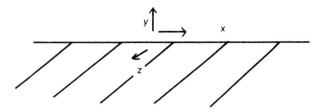

Fig. 12.8. Geometry for surface waves.

Let us also assume that the magnitude of the displacement of a solid element in any direction is a function of the depth. This means that we are assuming that

$$u_x = f_x(y)e^{i(kx-\omega t)},$$
$$u_y = f_y(y)e^{i(kx-\omega t)}, \qquad (12.E.2)$$
$$u_z = f_z(y)e^{i(kx-\omega t)}.$$

If we put these assumed forms of the solution back into the equation of motion, we will have

$$\frac{\partial^2 u_i}{\partial t^2} = c_i^2 \nabla^2 u_i, \qquad (12.E.3)$$

where $c_i = c_l$ if $i = x$, and c_t if $i = y, z$.

It must be emphasized that although c_l [given by Eq. (12.D.3)] and c_t [given by Eq. (12.D.6)] are velocities of body waves, they are not the velocity of any wave in the surface. They are simply different combinations of the parameters ρ, λ, and μ.

If we substitute the assumed forms of u_i into these equations, we find

$$\frac{d^2 f_i(y)}{dy^2} = \left(k^2 - \frac{\omega^2}{c_i^2}\right) f = \gamma_i^2 f(y), \qquad (12.E.4)$$

which has as its solution functions of the form $f = e^{\pm \gamma y}$. If we throw out solutions which become infinite as $y \to -\infty$, and note that we will prove later that $\gamma > 0$, we have

$$u_i = B_i e^{\gamma_i y} e^{i(kx-\omega t)}. \qquad (12.E.5)$$

This solution for u_x, u_y, and u_z exhibits all the properties we wish to associate with a surface wave—each component exhibits wave behavior, but as we go into the interior of the material, the disturbance dies out exponentially (but note that the three components do not die out at the same rate). However, as was implied in the introduction to this section, it is necessary to satisfy conditions at the boundary as well as the equation of motion if we wish to show that such waves exist.

The boundary condition, of course, is simply the requirement that the plane $y = 0$ be a free surface, which means that

$$\sigma_{yx}(y = 0) = \sigma_{yy}(y = 0) = \sigma_{yz}(y = 0) = 0. \tag{12.E.6}$$

Let us look at the condition on σ_{yz} first. From Eq. (12.C.1), we have

$$\sigma_{yz}(y = 0) = 2\mu \left[\frac{\partial u_y}{\partial z} + \frac{\partial u_z}{\partial y} \right]_{y=0} = 0, \tag{12.E.7}$$

but from symmetry, $\partial u_y / \partial z = 0$, so we have

$$\frac{\partial u_z}{\partial y} \bigg]_{y=0} = 0.$$

But from Eq. (12.E.2),

$$\frac{\partial u_z}{\partial y} \bigg]_{y=0} = \gamma_z B_z e^{i(kx - \omega t)},$$

so that we must have

$$B_z = 0. \tag{12.E.8}$$

In other words, in surface waves of the type we are studying there can be no displacement in the z-direction. In the language of seismology, displacement in the z-direction is called *SH* (for shear horizontal), since the displacement is horizontal to the surface in which the wave is propagating. On the other hand, *SV* (shear vertical wave) is one in which u_y is nonzero.

From the condition on σ_{xy}, we have [again using Eq. (12.C.1)]

$$\sigma_{xy}(y = 0) = 0 = \mu [B_x \gamma_x + ik B_y] e^{i(kx - \omega t)},$$

which means that

$$B_x = -\frac{ik}{\gamma_x} B_y. \tag{12.E.9}$$

If we put these back into Eq. (12.E.2), we find that at $y = 0$, we have

$$u_x^2 + \frac{k^2}{\gamma_x^2} u_y^2 = \text{const.}, \tag{12.E.10}$$

which means that the particle motion associated with this wave is in fact retrograde ellipse (see Fig. 12.9).

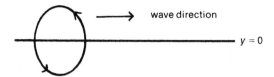

Fig. 12.9.   The motion of a particle in a surface wave.

This type of wave, which is confined to the surface and has no *SH* component, is called the *Rayleigh wave*, after Lord Rayleigh, who first discussed it.

What is the velocity of a Rayleigh wave? In order to answer this, we must apply the final boundary condition. From Eqs. (12.C.1), (12.E.2), and (12.E.6), we have

$$\sigma_{yy}(y = 0) = (ikB_x\lambda + (\lambda + 2\mu)\gamma_y B_y)e^{i(kx - \omega t)} = 0. \qquad (12.E.11)$$

Using Eq. (12.E.9) and rearranging, we have

$$\gamma_x\gamma_y = \frac{\lambda k^2}{\lambda + 2\mu},$$

which can be written, using Eqs. (12.E.4), (12.E.1), (12.D.3), and (12.D.6)

$$\left(1 - \frac{c^2}{c_l^2}\right)^{1/2}\left(1 - \frac{c^2}{c_t^2}\right)^{1/2} = 1 - \frac{2c^2}{c_t^2}. \qquad (12.E.12)$$

This equation determines c in terms of c_t and c_l (or, conversely, in terms of μ and λ).

Now we could, in principle, go ahead and solve this equation, inserting for c_t and c_l some quantities appropriate for the earth's surface. However, our job is made considerably simpler if we make use of an experimental observation known as *Poisson's relation*, which states that for the earth, it is approximately true that the Lamé coefficients are about equal. This, in turn, implies that

$$c_l^2 \approx 3c_t^2. \qquad (12.E.13)$$

In this case, the equation for c can be solved simply to give

$$\frac{c^2}{c_t^2} = 2 \pm \frac{2}{\sqrt{3}}. \qquad (12.E.14)$$

Which sign should we pick? To answer this question, we have to refer to Eq. (12.E.5), in which it was shown that a surface wave could exist in a solid. In order for this to be true, it was necessary that $c^2 < c_t^2 < c_l^2$. Thus, only the choice of the minus sign in Eq. (12.E.4) will result in a surface wave, and the other root must be discarded as extraneous. We are left with the result

$$c = 0.92c_t, \qquad (12.E.15)$$

so that the Rayleigh wave travels at a slightly slower velocity than the shear body wave.

In seismology, then, we expect that in addition to the two body waves discussed in Section 12.D, there will be a wave traveling along the surface of the earth as well. This means that in addition to the two signals discussed in the example of the Tokyo earthquake, a third signal will be received. This signal will arrive after the S and P signals (because it has a lower velocity and farther to travel), and will be pure SV in nature.

Such waves are, of course, observed in nature. In addition, it is also true that yet another kind of surface wave is observed, which is a pure SH wave. Although such a wave would not be possible in a uniform homogenous earth, they are possible in more realistic models of the earth, and it is to this problem we now turn our attention.

F. WAVES IN SURFACE LAYERS

The reason that we failed to predict the existence of SH surface waves in the previous section was that we had taken too simple a model for the earth. In actual fact, the earth is not a homogeneous medium, but has a rather complex structure. This will be discussed more fully in the next chapter, but for our purposes, we need only observe that a better model for the surface of the earth would be one in which there was a surface layer of a different material from the main body. The existence of waves in such a layer was first noted by A. E. H. Love, and they are usually called *Love waves*.

Suppose we take as our model of the earth's surface the situation shown in Fig. 12.10, where there is a semi-infinite solid extending from $y = 0$ downward, with density ρ' and Lamé coefficient μ', and a layer of solid of density ρ and coefficient μ from $y = 0$ to $y = T$.

If we again consider a wave moving in the x-direction, then an SH wave would correspond to a motion of the elements of the solids in the z-direction. Thus, we shall have to look for solutions to the equations of motion and the boundary conditions which are of the form

$$u_z = f(y)e^{i(kx-\omega t)},$$
$$u_y = u_x = 0, \qquad (12.F.1)$$

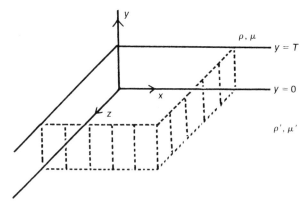

Fig. 12.10.   The geometry for Love waves.

in each of the two media. Following the steps that led to Eq. (12.E.4), we find that in each medium, we have an equation for $f(y)$ of the form

$$\frac{d^2 f}{dy^2} = \left(k^2 - \frac{\omega^2}{c_t^2}\right) f - \gamma^2 f, \tag{12.F.2}$$

where c_t is the shear wave velocity appropriate to each medium. By assumption, there is no longitudinal wave in this system.

As before, we need only consider $\gamma' > 0$ in the lower medium, since the motion must stay finite as $y \to -\infty$. In the surface layer, however, there is no restriction on γ, so that we have

$$f(y) = A \sin \gamma y + B \cos \gamma y \tag{12.F.3}$$

for the function $f(y)$ in the surface layer, and

$$f'(y) = E e^{\gamma' y} \tag{12.F.4}$$

for the function $f'(y)$ in the lower medium, and F, A, and B are unknown constants. The quantities γ and γ' differ in the value of c_t which appears in Eq. (12.F.2).

If this disturbance is to represent a physically realizable situation, three boundary conditions must be satisfied:

(i) the medium must be continuous at $y = 0$, which means

$$u_z(y = 0) = u_z'(y = 0), \tag{12.F.5}$$

(ii) the stresses must vanish at the free surface $y = T$, so that

$$\sigma_{yy}(y = T) = \sigma_{yx}(y = T) = \sigma_{yz}(y = T) = 0, \tag{12F.6}$$

(iii) the stresses must be continuous at the interface $y = 0$.

From condition (i), we immediately find that

$$E = B. \qquad (12.F.7)$$

It is easy to see that condition (ii) on σ_{xy} and σ_{yy} is trivially satisfied. The condition on σ_{zy} is just

$$\sigma_{yz}(y = T) = \mu \left. \frac{\partial u_z}{\partial y} \right]_{y=T} = 0,$$

so that

$$\tan \gamma T = \frac{A}{B}. \qquad (12.F.8)$$

Condition (iii) for σ_{yz} at the interface becomes

$$2\mu' \left. \frac{\partial u_{z'}}{\partial y} \right)_{y=0} = 2\mu \left. \frac{\partial u_z}{\partial y} \right)_{y=0},$$

which becomes, using Eq. (12.F.1),

$$\frac{A}{E} = \frac{\mu' \gamma'}{\mu \gamma} = \frac{A}{B}, \qquad (12.F.9)$$

where the second equality follows from Eq. (12.F.7).

Combining this result with Eq. (12.F.8), we find

$$\tan \gamma T = \frac{\mu' \gamma'}{\mu \gamma} = \frac{\mu'}{\mu} \left[\frac{1 - c^2/c_t'^2}{1 - c^2/c_t^2} \right]^{1/2}, \qquad (12.F.10)$$

which is an equation which relates $c = \omega/k$ to c_t, and hence determines the velocity of the Love wave, just as Eq. (12.F.12) determined the velocity of the Rayleigh wave. It should be noted, however, that unlike the Rayleigh wave, the Love wave will have a velocity dependent on the wavelength. This type of phenomenon was observed in surface waves in fluids in Chapter 5.

SUMMARY

The strain and stress tensors provide a general description of the deformation of a solid and the applied forces. They are related, for an elastic solid, by Hooke's law, although other kinds of relations are possible. Combining Hooke's law with Newton's second law led to an equation of motion for solids which, in turn, results in the existence of acoustic and shear waves in the body of a solid, and in Rayleigh waves in the surface. If we add the existence of a surface layer, a second kind of surface wave, the Love wave, is also seen to exist.

PROBLEMS

12.1. Write down Hooke's law in tensor form in Cartesian, cylindrical, and spherical coordinates. (*Hint*: You may find it useful to go back to the definition of u_{ik} in terms of a change in length.)

12.2. Show that the tensors Π_{ik} [defined in Eq. (1.C.11)] and σ_{ik} [defined in Eq. (8.A.9)] are stress tensors in the sense of Section 12.B.

12.3. In addition to the internal forces canceling, leaving only a surface force, as discussed in Section 12.B, the internal moments in a solid must do the same.

(a) Show that the total moment in a solid can be written

$$M_{ik} = \int \left[\frac{\partial \sigma_{il}}{\partial x_l} x_k - \frac{\partial \sigma_{kl}}{\partial x_l} x_i \right] dV.$$

(b) Show that this can be converted to a surface integral, except for a term of the type

$$\int (\sigma_{ik} - \sigma_{ki})\, dV.$$

(c) Hence give an argument that the stress tensor must be symmetric.

12.4. In Section 12.C, we discussed the idea of a Newtonian solid. In such a solid, Hooke's law is replaced by an equation in which the stress is proportional to the time derivative, or rate, of the strain, rather than to the strain itself. Using arguments analogous to those leading to Eq. (12.C.1), write down the equation relating stress and strain for such a solid. Show that in the case of an incompressible solid, this becomes

$$\sigma_{ik} = \eta \frac{\partial u_{ik}}{\partial t},$$

where η is a constant.

12.5. Show that for an incompressible elastic solid

$$\mu = \frac{E}{2},$$

where E is Young's modulus and μ is the Lamé coefficient.

12.6. When will the Rayleigh wave arrive at San Francisco in the example in Section 12.D?

12.7. In Chapter 8, we defined viscosity in terms of a stress tensor and an argument based on Occam's razor. An alternate way of defining viscosity is as follows: Consider a cylindrical tube with a fluid flowing in the z-direction. Take an element of the fluid and show that Newton's second law is

$$\rho \frac{Dv}{Dt} = -\frac{\partial p}{\partial z} - \frac{1}{r}\frac{\partial}{\partial r}(r\sigma_{rz}),$$

where σ_{rz} is the internal stress generated by the fluid motion. If we define the coefficient of viscosity by

$$\sigma_{rz} = -\eta \frac{\partial v}{\partial r},$$

show that we recover the Navier–Stokes equation.

12.8. Consider a circular cylinder which is being twisted by a force F applied tangentially. Let the result of the force be that an element at the edge is moved through an angle ϕ, as shown. Show that the torque on the cylinder is related to the angle by

$$T = \frac{1}{2} \frac{\pi \mu \phi a^4}{l}.$$

The quantity Tl/ϕ is called the *torsional rigidity*.

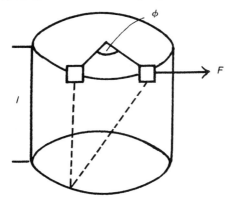

What is the energy stored in the twisted cylinder?

12.9. It should be obvious that Eq. (12.C.16) can be applied in the case where no motion is present in the solid (i.e. for the static cases treated in the previous chapter). To make this point, consider a solid sphere of inner and outer radii a and b, respectively, with pressure P_1 inside and P_2 outside.

(a) From the symmetry of the problem, show that the equation for the displacement is just

$$\nabla(\nabla \cdot \mathbf{u}) = 0.$$

(b) Hence show that the strain tensor is given by

$$u_{rr} = A - \frac{2B}{r^3}$$

and

$$u_{\theta\theta} = A + \frac{B}{r^3},$$

where A and B are as yet undetermined constants.

(c) From part (b), show that the radial component of the stress tensor is

$$\sigma_{rr} = \frac{E}{1 - 2\sigma} A - \frac{2EB}{(1 + \sigma)} \frac{1}{r^3},$$

where

$$A = \frac{(P_2 b^3 - P_1 a^3)}{a^3 - b^3} \frac{(1 - 2\sigma)}{E}$$

and

$$B = \frac{a^3 b^3 (P_2 - P_1)}{b^3 - a^3} \frac{(1 + \sigma)}{2E}.$$

12.10. From the results of Problem 12.9, derive the stresses for the following two limiting cases:

(a) A thin spherical shell of thickness h and radius R surrounded by a vacuum and maintaining a pressure P inside.

(b) A spherical cavity in an infinite medium, with a pressure P inside of it. Can you think of any applications where these limits might be useful?

12.11. Consider the case of a solid which is undergoing a plane deformation: i.e. a deformation in which $u_z = 0$ everywhere in the solid.

(a) For the static case, show that the equations of motion can be reduced to two equations in the stress tensor.

(b) Define a function χ by the relations

$$\sigma_{xy} = \frac{\partial^2 \chi}{\partial y^2}, \qquad \sigma_{xy} = -\frac{\partial^2 \chi}{\partial x \partial y}, \qquad \sigma_{yy} = \frac{\partial^2 \chi}{\partial x^2},$$

and show that these forms of the tensor automatically satisfy the equations in part (a).

(c) Hence show that the function χ, called the *stress function*, must satisfy the equation

$$\nabla^2 (\nabla^2 \chi) = 0,$$

which is called the *biharmonic equation*.

12.12. In Section 12.C, we derived the equations of motion in terms of the stress tensor. This is, of course, the most usual and useful form of this equation. There is another form, however, which can be written solely in terms of the strain tensor. Use Hooke's law to write it down.

12.13. Consider a cylinder of radius a rotating with frequency ω about its axis of symmetry.

(a) Write down the equation of motion for such a system.

(b) Show that the solution to the equation is

$$u_r(r) = \frac{\rho \omega^2 (1 + \sigma)(1 - 2\sigma)}{8E(1 - \sigma)} r[(3 - 2\sigma)a^2 - r^2].$$

12.14. In the text, we consider elastic waves in infinite or semi-infinite media only. Let us ask what happens when we consider waves in thin rods or sheets of the type considered in Chapter 11.

(a) Consider a longitudinal wave in a thin rod. Let the wave travel in the z-direction. Show that the velocity of the wave in this case is

$$c_l = \sqrt{\frac{E}{\rho}}.$$

How does this compare to waves in an infinite medium?

(b) Consider a longitudinal wave traveling in a thin plate in the z-direction. Show that in this case we find

$$c_l = \sqrt{\frac{E}{\rho(1 - \sigma^2)}}$$

for the velocity of the disturbance associated with u_z.

12.15. Consider a beam of the type discussed in Section 11.C which can be bent, but need no longer be in static equilibrium.

(a) Show that the equation of motion for such a system is

$$q\frac{\partial^2 y}{\partial t^2} = EI\frac{\partial^4 y}{\partial x^2}.$$

(b) Determine the frequencies at which the rod may vibrate, if it is clamped at one end and free at the other.

(c) Hence show that the smallest frequency is

$$\omega_{min} = \frac{3.52}{L^2}\sqrt{\frac{ET}{q}},$$

where q is the mass per unit length of the rod. (*Hint*: Assume that the solutions of the equation are separable, and that the integration of the $X(x)$ equation is a sum of trigonometric and hypertrigonometric functions of x.) This is the theory of the tuning fork.

REFERENCES

K. E. Bullen, *An Introduction to the Theory of Seismology*, Cambridge U.P., 1965.
   An excellent modern discussion of the theory of solids as applied to seismology. The mathematics level is fairly high.
Sir Harold Jeffreys, *The Earth*, Cambridge U.P., 1970.
   One of the classic texts in geophysics. Very complete and readable, with many examples and several sections on the origin of the earth.

For related reading, see

J. D. Jackson, *Classical Electrodynamics*, John Wiley and Sons, New York, 1972.
   This text contains a discussion of the Maxwell stress tensor in Chapter 6.

13

Applications of Seismology: Structure of the Earth and Underground Nuclear Explosions

A vast, limitless expanse of water ...
spread before us. ... 'The Central Sea' said my uncle

JULES VERNE
A Journey to the Center of the Earth

A. SEISMIC RAYS

In the previous chapter, the existence of waves in elastic solids was discussed and the way in which these waves could be used to gain information about the structure of the earth was hinted at. All of this discussion (except for the development of Love waves) was done in the context of a uniform earth of constant density and Lamé coefficients. We know, of course, that the earth is really not so simple as that, and one of the main goals of seismology is to try to discover the details of the structure of the earth. The problem can be put in the following way: Given that we can only make measurements at the surface of the earth, what can we do to discover the structure of the interior?

One obvious way to answer this question is to try to measure something that passes through the interior, and is affected by it. This is somewhat analogous to a physician taking X-rays of the human body, and learning from the absorption of the radiation the condition in the interior.

The only "radiation" of this type that we have at our disposal are the waves, discussed in the last chapter, which can propagate in solids. Obviously, the surface waves will be of limited value in exploring the deep interior of the earth, and we will confine our attention to body waves for the moment. Let us reconsider the example of Section 12.D in which

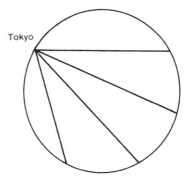

Fig. 13.1   Seismic rays for a Tokyo earthquake.

an earthquake occurred in Tokyo. If the earth were uniform, the seismic waves would propagate out from Tokyo, and could be measured at many places on the earth (see Fig. 13.1). At each observing station, the time of the arrivals of the P and S waves could be measured. In this problem, there are three numbers which we do not know, but would like to learn. These are the density of the earth and the two Lamé coefficients. Thus, if we could measure three different time intervals (corresponding to three different observation stations), we could completely determine ρ, λ, and μ. While data from a single observation point cannot tell us much about the interior of the earth, data from many stations, taken together, can do so quite well. This, incidentally, is the reason that international cooperation has been so important in the development of seismology.

The earth, of course, is not uniform. The real problem of seismology is to discover the density and elastic properties of the materials inside of the earth as a function of depth and position. This means that instead of trying to fix three unknown constants, as in the above example, the seismologist is actually trying to fix densities and elastic constants (and deviations from elasticity) as a function of depth in the earth. As a start toward solving this problem, let us ask how a seismic ray propagates through a medium which is not uniform.

To do this, let us consider what happens when a plane wave (either S or P) arrives at a boundary between two layers, each characterized by a different velocity (which, in turn, is related to different elastic constants). We can see what will happen to the wave by invoking *Huygens principle*, familiar from optics, which tells us that each point of a wave front can be thought of as emitting a spherical wavelet with the wave at any other

Fig. 13.2(a). The propagation of a wave by Huygens wavelets.

point being given by the sum of the wavelets. For example, a plane wave propagating in a uniform medium could be thought of as shown in Fig. 13.2(a), with each point of the front emitting a wavelet, and these wavelets adding up to give the wave front farther downstream.

When such a wave encounters a boundary, however, the situation will be as pictured in Fig. 13.2(b). The wavelets emitted from the point P will travel at a velocity v', characteristic of the second medium, while those emitted at Q will continue to travel with velocity v. If it takes time t for the wavelet from Q to travel to the interface, then the wavelet from P will have traveled a distance $v't$. Hence the new wave front will be the line AB, and we see that the net effect of the interface is to change the direction of the wave. This phenomenon, known as refraction, is familiar in optics. From the geometry in Fig. 13.2(b), it is easy to see that

$$\frac{\sin \theta}{v} = \frac{\sin \theta'}{v'}, \tag{13.A.1}$$

which is the familiar *Snell's law* for refraction.

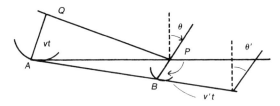

Fig. 13.2(b). Geometry for the derivation of Snell's law.

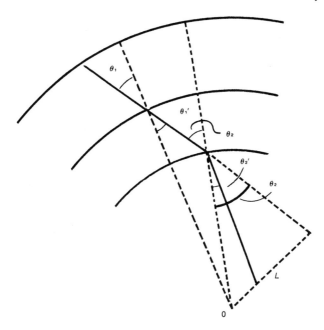

Fig. 13.3. Propagation of a seismic wave in a layered medium.

Let us now consider the interior of the earth as a series of layers, as shown in Fig. 13.3. Then a wave which starts off at an angle θ_1 will be successively refracted at each interface, with the relationship

$$\frac{\sin \theta_1}{v_1} = \frac{\sin \theta_1'}{v_2}, \qquad \frac{\sin \theta_2}{v_2} = \frac{\sin \theta_2'}{v_3}$$

following from Snell's law. By geometry, however, we have

$$L = r_1 \sin \theta_1' = r_2 \sin \theta_2.$$

From the above two equations, we see immediately that

$$\frac{r_1 \sin \theta_1}{v_1} = \frac{r_2 \sin \theta_2'}{v_3} = \frac{r_1 \sin \theta_1'}{v_2} = \frac{r_2 \sin \theta_2}{v_2}.$$

The extension of this type of relationship to an infinite number of layers (which would represent a continuously changing interior) yields the general law

$$p = \frac{r \sin \theta}{v}, \qquad (13.A.2)$$

where p is a constant along the entire ray (it is called the ray parameter). It follows that waves traveling through the earth do not, in fact follow straight lines as in Fig. 13.1, but curves, as in Fig. 13.4.

There are several points which should be made before proceeding. First, it should be obvious that, in general, S and P waves starting from the same point will have different paths in the interior, since, in general, the dependences of c_t and c_l on r will not be the same. Second, in addition to the phenomenon of refraction in the earth, seismic waves (like any other waves) can be reflected at interfaces as well. This will be shown in Problems 13.3, 13.4, and 13.5.

The general problem faced by the seismologist, then, is to understand the relation between the time and place of arrival of a seismic wave, and the trajectory which it has followed through the earth. Let us examine this problem in more detail. We know that each seismic wave is characterized by a parameter p, and travels through the earth subtending an angle Δ at the center, taking a time T to get from P_0, the point of emission, to Q_0, the point of detection (see Fig. 13.4).

If we denote by s the distance along the curve P_0Q_0, then by simple geometry Eq. (13.A.2),

$$p = \frac{r \sin \theta}{v} = \frac{r}{v} \cdot \frac{r \, d\alpha}{ds}, \tag{13.A.3}$$

but in general,

$$ds^2 = dr^2 + r^2 \, d\alpha^2,$$

so that, if we let $\eta = r/v$,

$$d\alpha = \frac{p}{r(\eta^2 - p^2)^{1/2}}. \tag{13.A.4}$$

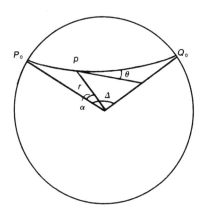

Fig. 13.4. The path traversed by a seismic ray.

If we integrate from the point of emission, P_0, to the halfway point along the trajectory (at $r = r_1$), we have

$$\frac{1}{2}\Delta = p \int_{r_1}^{r_0} \frac{dr}{r(\eta^2 - \rho^2)^{1/2}} \tag{13.A.5}$$

for the angle subtended by the trajectory. In a completely analogous way, we can derive an expression for the transit time

$$\frac{1}{2}T = \int_{r_1}^{r_0} \frac{\eta^2 \, dr}{r(\eta^2 - \rho^2)^{1/2}} \tag{13.A.6}$$

from the fact that $ds = v \, dt$.

These equations relate the angle Δ and time T, both of which are measurable quantities, to integrals involving $v(r)$ and the ray parameter p. But since

$$p = \frac{r_0 \sin \theta}{v}\bigg]_{P_0} = \frac{r_0 \sin \theta}{v}\bigg]_{Q_0}, \tag{13.A.7}$$

p can also be determined by surface measurements. Thus, by measuring arrival times of waves at different points about the earth, we can determine $v(r)$ in the interior. In actual practice, there is more data than just seismic arrival times. We have also the free vibrations of the earth (see Chapter 7) and some good theoretical conjectures about the chemical composition of the interior which must be fit into the results as well.

The general picture of the earth's interior which has arisen from such studies is illustrated in Fig. 13.5. The outer layer of the earth, the crust, is only about 15 km thick. Under the crust is a solid mantle, which is itself

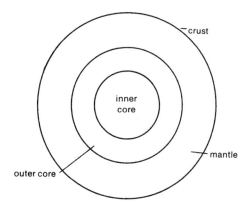

Fig. 13.5.   The general structure of the earth.

usually divided into upper, middle, and lower regions. The mantle extends to a depth of about 2800 km. Between the mantle and the crust is a sharp transitional region known as the *Mohorovičič discontinuity.* We believe that the continents, which are part of the crust, actually float on the mantle, and have moved around during geological times. The subject of *continental drift* is a fascinating one, and one of the more important ideas of modern geophysics.

Below the mantle, there is the core. The outer core, extending down to about 5000 km, is liquid metal, and it is thought that the motions of this liquid core are important in generating the magnetic field of the earth. At the very center of the earth is the inner core, composed of solid metals.

Thus, we see that one application of the theory of elasticity is to give us an increasingly detailed picture of the earth on which we live. Aside from the obvious practical advantages of such knowledge, this also gives us important information about the process by which the earth, and hence the solar system, were formed.

B. UNDERGROUND NUCLEAR EXPLOSIONS

Another application of the knowledge of waves in solids is in the field of arms control. The ability to limit the development of nuclear weapons depends directly on the ability to detect nuclear tests. When such tests are carried out in the atmosphere, the detection is relatively simple, since prevailing winds will carry radioactive debris across national boundaries to detecting stations. Underground tests, however, are not so easy to detect, since the debris is confined (barring an accidental release of radioactive materials into the atmosphere). In fact, the only indication that such a test has occurred which would be detectable at large distances from the site of the test would be the seismic signal generated by the explosion. This, in turn, leads us to the question of how seismic waves are generated.

Before turning to this question, however, let us review briefly the sequence of events which follows a nuclear explosion. Immediately following the blast, tremendous pressure (on the order of 10^6 atmospheres) are present. The sudden release of energy completely strips the atoms in the neighborhood of the blast, and two things occur: (1) a burst of electromagnetic radiation moves away from the blast site, and (2) the debris of the blast moves away also, forming a shock front. At the beginning, the radiation front moves quickly, heating up the surrounding material and forming an expanding "fireball" of high temperature gases. As the fireball expands, its temperature drops (why?) and the expansion slows down. At some point, called *breakaway*, the shock wave overtakes the radiation front and moves ahead of it.

In atmospheric explosions, this is a complete description of the blast phenomenon. In underground explosions, however, there is another quantity which enters and that is the size of the cavity in which the explosion occurs. For the sake of simplicity, we will assume throughout the rest of this section that we are dealing with a spherically symmetric geometry. If the radius of the cavity is less than the radius at which breakaway occurs, then the fireball will actually strike the cavity walls, vaporizing them. Since more energy is required to vaporize rock than to heat up air, the fireball will be slowed down. When the shock front catches up with the fireball and moves ahead, one of two things may happen: (i) the shock front will have sufficient energy to continue vaporizing the rock, (ii) the shock front will have only enough energy to melt the surrounding rock. In either case, as the shock wave proceeds out from the blast site the damage which it does decreases. At large distances, the rock will be fractured, but it is clear that at some distance, which we shall denote by R_σ (the "seismic" radius), the deformation of the rock caused by the shock front will not exceed the elastic limits, and the rock will simply be deformed elastically, which means that it will exert internal forces which will bring it back to its original position. We speak of the shock wave "decaying" into an elastic wave at this point. The question which we must ask has to do with relating the deformation at R_σ to the seismic wave which would be detected at large distances.

It should be obvious from the foregoing discussion that it is possible to heighten or reduce the effects of the blast at R_σ by choosing the cavity radius to be greater or less than the fireball radius, and by choosing the material surrounding the blast site. Thus, a small cavity in solid rock (a "tamped" explosion) would produce much greater seismic signals than a large cavity in a very porous material. This problem, which involves the *coupling* of the explosion to seismic waves, is obviously of great interest to those concerned with arms control. A much more detailed discussion is given in the text by Rodean (1971) cited at the end of the chapter.

The problem of detecting an underground test, then, becomes one of understanding what sort of seismic signals such a test would generate. Let us consider a spherically symmetric situation such as that in Fig. 13.6, in which some known displacement of the material takes place at $r = R_\sigma$, and waves propagate out. We know that the equations which govern the displacements of the solid at large radii are

$$\frac{1}{c_l^2} \frac{\partial^2 u_r}{\partial t^2} = \nabla^2 u_r = \frac{1}{r^2} \frac{\partial}{\partial r} \left(r^2 \frac{\partial u_r}{\partial r} \right) \tag{13.B.1}$$

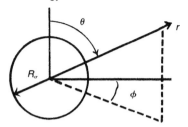

Fig. 13.6.   Coordinates for the underground nuclear explosion.

for displacements in the r-direction, and

$$\frac{1}{c_t^2} \frac{\partial^2 u_\theta}{\partial t^2} = \nabla^2 u_\theta = \frac{1}{r^2} \frac{\partial}{\partial r} \left(r^2 \frac{\partial u_\theta}{\partial r} \right) \tag{13.B.2}$$

for displacements in the θ-direction. A similar equation can be written for the ϕ-direction, of course.

If we make the substitution

$$\chi_r = \frac{u_r}{r}, \tag{13.B.3}$$

then Eq. (13.B.1) becomes

$$\frac{\partial^2 \chi_r}{\partial r^2} = \frac{1}{c_l^2} \frac{\partial^2 \chi_r}{\partial t^2}, \tag{13.B.4}$$

which is just the wave equation. Without loss of generality, we will consider only plane wave solutions, so that

$$u_r(r, t) = \frac{A}{r} e^{ik(r - c_l t)}. \tag{13.B.5}$$

By exactly similar steps, we would have

$$u_\theta(r, t) = \frac{B}{r} e^{ik(r - c_t t)}. \tag{13.B.6}$$

In order to determine the constants A and B, it is necessary to refer to the boundary conditions at the point $r = R_\sigma$. We know that at this point there is no external force on the rock until the time of the explosion, then forces are applied to the material, and these forces will die out gradually a long time after the explosion. In general, the applied force at the seismic radius would look like the one shown in Fig. 13.7. Whatever the actual functional dependence of the force, however, it is clear that we can always write

$$F(t) = \sum_n a_n e^{i\omega_n t}; \qquad \omega_n = \frac{2\pi n}{T},$$

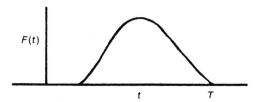

Fig. 13.7.   A typical applied force at the seismic radius.

so that we can, for the sake of our problem, consider only the Fourier component

$$F(t) = F_0 e^{i\omega_n t}.$$

We now ask the critical question. In which direction is this force pointed? For an underground explosion, we would expect the force to be mainly radial, so that the seismic radius,

$$\sigma_{rr}(t) = F(t),$$
$$\sigma_{r\theta} = \sigma_{r\phi} = 0. \tag{13.B.7}$$

For an earthquake, or other natural source of the seismic signal, on the other hand, we would expect that $\sigma_{r\theta}$ and $\sigma_{r\phi}$ would not vanish at $r = R_\sigma$. This, then, is the main difference between underground explosions and naturally occurring events. We must now see what effect this difference in boundary conditions will have on seismic signals far from the event.

From Hooke's law for the case of spherical symmetry, we have

$$\sigma_{rr} = (\lambda + 2\mu)\frac{\partial u_r}{\partial r} + 2\mu \frac{u_r}{r}, \tag{13.B.8}$$

so that at the seismic radius, combining Eqs. (13.B.8), (13.B.7), and (13.B.5), we have

$$\sigma_{rr}(r = R_\sigma) = A e^{ikc_l t} \frac{e^{ikR_\sigma}}{R_\sigma}\left[(\lambda + 2\mu)ik - \frac{\lambda}{a}\right]$$
$$= F_0 e^{i\omega_n t}, \tag{13.B.9}$$

which leads immediately to the result

$$k = \frac{\omega_n}{c_l}$$

and

$$A = F_0 R_\sigma e^{-ikR_\sigma}\left[ik(\lambda + 2\mu) - \frac{\lambda}{R_\sigma}\right]^{-1}. \tag{13.B.10}$$

Thus, the amplitude of the P wave at a large distance from the source is directly proportional to the magnitude of the applied forces. The S wave, on the other hand, must be determined from the requirement that at $r = R_\sigma$,

$$\sigma_{r\theta} = \mu \left[\frac{\partial u_\theta}{\partial r} - \frac{u_\theta}{r} \right]_{r=R_\sigma} = 0, \qquad (13.\text{B}.11)$$

which leads immediately to the result

$$B = 0. \qquad (13.\text{B}.12)$$

Thus, in our simplified model, the signal characteristic of an underground explosion would be a missing S wave. Of course, in a real situation, the applied force would never be exactly radial, and some S wave would be generated. Nevertheless, a sharp diminution of S wave is one commonly accepted criterion for discriminating between small earthquakes and underground tests.

A more important tool, which we shall not discuss in detail, arises from the fact that the outgoing seismic waves from an underground event will strike the surface near the event and generate Rayleigh surface waves. Although the theory of how Rayleigh waves are generated in this manner is not really well worked out, it does turn out that sources which generate both S and P waves are much more efficient in creating Rayleigh waves at a free surface than are sources generating only P waves. This means that a second consequence of Eq. (13.B.12) is that in addition to the absence of the S body waves (an absence which is somewhat difficult to detect for small explosions with present techniques), there should be a great reduction in surface waves as well. This has, in fact, been observed, and is discussed in some of the references at the end of the chapter.

We see then, that a relatively simple model of the seismic response to an underground explosion can explain some of the ideas which are now being examined in research on nuclear arms control.

SUMMARY

We have seen how the knowledge about waves in solids could be applied to two separate problems. First, we saw that body waves traveling through the earth would follow trajectories which depended on the elastic constants in the interior. This becomes then a method of finding out about the structure of the interior of the earth.

Second, we saw that underground nuclear explosions and earthquakes are quite different as far as the type of seismic waves which they generate are concerned. An explosion would be expected to have much smaller S waves and surface waves than an earthquake.

PROBLEMS

13.1. For the example of the Tokyo earthquake of Section 12.D, construct a table of time intervals between the event and the arrivals of the S and P waves at 10 different points around the world (you may choose your own points), assuming a uniform earth.

13.2. Consider a ray starting at P_0 and ending at Q_0, as in Fig. 13.4, and let T and Δ be the travel time and subtended angle for this ray, and p be the ray parameter. If a ray starts from a neighboring point, and has $T + dT$, $\Delta + d\Delta$, and $p + dp$ for the corresponding values, show that

$$p = \frac{dT}{d\Delta}.$$

13.3. Consider a free surface at $z = 0$ with a P wave incident with angle θ. Take the incident wave to be of the form

$$\psi = A e^{i[k(x - z \tan \theta) - \omega t]}.$$

(a) Write down the boundary conditions at the surface $z = 0$.

(b) Assume that there will be both a reflected P wave and a reflected S wave, and take their form to be

$$\psi_P = A_1 e^{i[k(x + z \tan \theta) - \omega t]}$$

and

$$\psi_S = B_1 e^{i[k(x + z \tan \phi) - \omega t]}.$$

Show that it is not possible to satisfy the boundary conditions if $B_1 = 0$ so that there *must* be a reflected S wave.

(c) From the equations of motion, show that

$$\frac{\cos \theta}{c_l} = \frac{\cos \phi}{c_t}.$$

(d) Show that the coefficients of the reflected wave are given by

$$\frac{A - A_1}{A + A_1} = \frac{(\tan^2 \phi - 1)^2}{4 \tan \phi \tan \theta},$$

$$A + A_1 = \frac{2 \tan \phi}{\tan^2 \phi - 1} B_1.$$

13.4. (a) Show that if an SV wave were incident on the surface in Problem 13.3, of magnitude B, and the reflected P and S waves had amplitude A_1 and B_1, respectively, that retracing the steps in Problem 13.3 would give

$$\frac{B - B_1}{B + B_1} = \frac{(\tan^2 \phi - 1)^2}{4 \tan \phi \tan \theta},$$

$$B + B_1 = \frac{-2 \tan \phi}{\tan^2 \phi - 1} A_1,$$

and that the result of part (c) still follows.

(b) Hence show that if $\cos \phi > c_t / c_l$, the reflected P wave will die out rapidly as we leave the surface, and the amplitudes of the incident and reflected S wave will be equal.

(c) Show that for an incident SH wave, the reflected wave is always equal in amplitude to the incident wave, and no P wave is generated at the surface.

13.5. Consider now a wave incident from below on an interface at $z = 0$, with the material in the lower half plane characterized by Lamé coefficient μ and λ, and the material in the upper half plane characterized by μ' and λ'. Assume the amplitudes of the waves are as follows:

A incident P wave,
B incident SV wave,
A_1 reflected P wave,
B_1 reflected SV wave,
A' transmitted P wave,
B' transmitted SV wave,
C incident SH wave,
C_1 reflected SH wave,
C' transmitted SH wave,

and assume that the angles associated with the directions of the transmitted P and S waves are θ' and ϕ', respectively.

(a) Write down the equations of motion in each medium and the conditions which must hold at $z = 0$.

(b) Show that the equations for the SH wave are independent of the equations for the P and SV waves (as was seen in Problems 13.3 and 13.4 above), and that

$$C + C_1 = C',$$

$$\mu \tan \phi (C - C_1) = \mu' \tan \phi' C'.$$

(c) Derive Snell's law for refraction from the boundary conditions in part (a).

(d) Write down the four (rather complicated) equations which determine A_1, B_1, A' and B'.

13.6. A liquid can be characterized by the statement that $\mu = 0$. Given the results of Problem 13.3, can you explain why no S waves are observed directly opposite an earthquake, although P waves are?

13.7. A rough parameterization of the velocity of seismic waves as a function of depth, which is useful in calculations, is

$$v = ar^b,$$

where a and b are constants. For the special case $b = 1$, consider a signal originating at a latitude Θ. The signal is observed at a point Q_0, at latitude Φ. Find the deepest penetration of the ray as a function of Φ.

13.8. Derive the equation analogous to Eqs. (13.B.10) and (13.B.12) for an earthquake in which the boundary conditions are

$$\sigma_{rr} = \sigma_{r\theta} = \sigma_{r\phi} = F(t)$$

at some radius R_σ. Is this a reasonable model of an earthquake?

REFERENCES

All of the geophysics texts cited in Chapter 12 contain discussions of seismic waves.

Bruno Rossi, *Optics*, Addison-Wesley, Reading, Mass., 1957.
    Chapter 1 contains an excellent description of Huygens principle applied to light.
F. D. Stacey, *Physics of the Earth*, John Wiley and Sons, New York, 1969.
    A descriptive, mainly nonmathematical discussion of seismology, the earth's magnetic and gravitational fields, and the internal structure of the earth.
R. H. Tucker, A. H. Cook, H. M. Iyer, and F. D. Stacey, *Global Geophysics*, American Elsevier, New York, 1970.
    A descriptive book covering seismology and the earth's structure.
F. G. Blake, Jr., "Spherical Wave Propagation in Solid Media," *J. Acoustical Soc. America* **24**, 211 (1952).
    A concise description of wave propagation.
B. Gutenberg, *Physics of the Earth's Interior*, Academic Press, New York, 1959.
    Detailed discussion of the layers and regions of the interior.
S. K. Runcorn (ed.), *Continental Drift*, Academic Press, New York, 1962.
    Collection of papers on all phases of the problem of continental drift.
Physics Today, March 1974.
    An entire issue devoted to discussions of modern ideas in geophysics.
H. C. Rodean, "Nuclear-Explosion Seismology," *U.S.A.E.C. Technical Information Bulletin* (TID-25572), 1971.
    Detailed description of effects of underground explosions, the problem of coupling, and the generation of seismic waves. Extensive bibliography.
H. R. Myers, "Extending the Nuclear Test Ban," *Scientific American* **226**, 13 (1972).
    A good review article on the present status of our abilities to detect underground nuclear explosions. No mathematics.

14

Applications to Medicine: Flow of the Blood and the Urinary Drop Spectrometer

> Physics is love, engineering is marriage.
>
> NORMAN MAILER
> *Of a Fire on the Moon*

A. INTRODUCTION

Throughout the text up to this point, we have concerned ourselves with more or less "conventional" applications of the physics of fluids and solids to areas of basic research in such fields as astronomy and geophysics, touching only briefly and occasionally on topics which might be considered "applied physics." Yet it is clear that with so much of the world around us composed of materials which are approximately ideal fluids or solids, one of the prime reasons for studying the subjects in this text is in order to be able to apply the simple physical principles which we have learned to real situations.

Perhaps in no area is this more true than in the area of the applications of physics to medicine and to an understanding of the human body. The body is, after all, a system which operates according to the same physical laws as do other natural systems. There are many parts of the body where it seems obvious that a simple physical model would explain a great deal of the observed behavior. The skeleton, for example, can be thought of as a structural system in which external loads are counteracted by internally generated forces, just as was the case for mountain chains in Chapter 11.

There are many fluid systems in the body, the most obvious of which is the circulatory system. But even at the cellular level classical processes of osmosis and diffusion through membranes are extremely important.

In this chapter, we shall discuss some simple models for two physical systems: The flow of blood through an artery, and the behavior of the external urine stream. The first of these is an old problem which has received a great deal of attention in the past, while the second represents a relatively new application of physical reasoning to diagnostic medicine.

The circulatory system can, with a great deal of oversimplification, be considered as shown in Fig. 14.1.

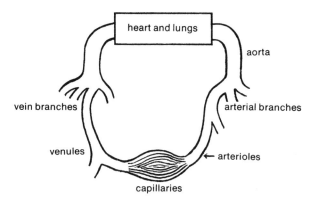

Fig. 14.1.   A schematic view of the circulatory system.

The blood is pumped from the heart and lungs through a system of branching arteries, whose size diminishes with distance from the heart. Eventually, it flows through the network of capillaries and back into the venous system, which returns it to the heart and lungs.

The basic problem of blood flow can be stated as follows: Given the time dependence of the pressure and the flow at the exit of the heart, and given the composition and layout of the arterial and venous systems, what will the flow and pressure be at any point in the body? This is an extremely complicated problem, and we are a long way from being able to describe the circulatory system mathematically. Perhaps a few remarks about the complexity of the system will help the reader to understand why.

The first problem is the nature of blood itself. Strictly speaking, it is not a fluid in the classical sense in which we have used the term up to this point, but is a suspension of small particles in a fluid (known as the plasma). The most important of these particles from the point of view of the circulation are the red blood cells, which are roughly the shape of a doughnut with the center partially filled in, and are typically about 7 microns across (1 micron = 10^{-4} cm). When we are dealing with arteries, whose dimensions are typically in the millimeter range, this is not too important an effect, but 5–10 μ is the size of a typical capillary. This means that flow in the artery will be quite different in character from that in a capillary. In the former, the size of the vessel is very large compared to the size of the cells, so that it is reasonable to treat the blood as a classical fluid. In capillaries, however, the cells must go through one at a time (the process is similar to pushing a cork through a bottle neck). In vessels of intermediate size, like the arterioles, the problem is even more complex.

Even if we restrict our attention to the arteries, we immediately encounter difficulties which we have not run into before. We have always argued that it is a good approximation to treat liquids as incompressible, so that the equation of continuity takes on a particularly simple form. In addition, we have always been able to assume that the coefficient of viscosity of a fluid, as defined in Eq. (8.A.9) was a constant, independent of the motion of the fluid. Because of its peculiar composition, neither of these assumptions is true for blood. It is, in fact, a relatively compressible fluid, and its coefficient of viscosity depends markedly on the velocity. This means that the Navier–Stokes equation becomes extremely complicated even if we can treat blood as a classical fluid, and explains the relatively primitive state of the theory of blood flow.

A second important complication in the problem of blood flow is the fact that the boundary conditions are no longer of the simple form we have grown accustomed to. The walls of the arteries are not rigid, but are in fact deformable solids. Thus, when a pulse comes down the artery, the walls expand. Nor is the arterial wall necessarily of the simple type which can be described by Hooke's law for an elastic material. In fact, the arterial wall is composed of a rather complicated material whose properties under stress fall into the general class of materials called *viscoelastic*. This means that the response to an applied force depends on the rate at which that force is applied, as would be appropriate for a Newtonian solid (see Section 12.C), as well as the usual restoring force which is proportional to the magnitude of the applied force. In addition, at large deformations, the structure of the arterial wall itself comes into play. It is composed of two substances, elastin and collagen (a third

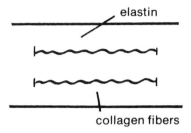

Fig. 14.2(a).   The arterial wall at rest.

structural component, smooth muscle, is not thought to have much effect on the elastic properties of the wall). The elastin is a rubbery, extensible material, while the collagen is more like a fiber which has a high resistance to deformation. The collagen is strung very loosely in the wall, with a lot of slack (see Fig. 14.2(a)), so that for small deformations, it has no effect on the walls. When the wall is stressed so that the slack is taken up, however, we have the situation in Fig. 14.2(b), in which the collagen now takes over and prevents further deformation of the artery. The biological usefulness of such a system is obvious, but equally obvious is the fact that such a structure is extremely difficult to describe mathematically.

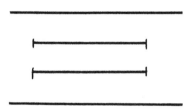

Fig. 14.2(b).   The arterial wall under tension.

Nevertheless, it is the job of the scientist to deal with complicated systems when they occur in nature. The general line of attack which is usually followed is to make a series of approximations which simplify the problem to the point where it is mathematically tractable, and then hope that the solution which is obtained has the main features of the system which we are trying to describe.

We shall follow this line in dealing with the problem of blood flow in the arteries. We shall begin by discussing the response of an elastic arterial wall to pressure, both static and time dependent, and then turn to the full problem.

B.   RESPONSE OF ELASTIC ARTERIAL WALLS TO PRESSURE

We shall begin by examining a problem whose applicability to the flow of blood in an artery is obvious. Consider a cylindrical elastic tube which contains some fluid whose pressure (not necessarily constant) is known. For the moment, we assume the pressure does not vary along the length of the tube. How does the tube respond?

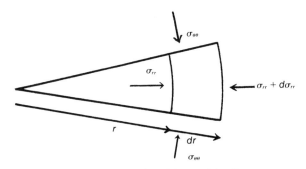

Fig. 14.3.   A section of the arterial wall.

Let us assume that there is azimuthal symmetry, and consider one infinitesimal volume element in the material. The stresses which act on this element are shown in Fig. 14.3. If we compute the forces in the r-direction, we have

$$F_r = - [\sigma_{rr}(r \, d\theta \, dz) + (\sigma_{rr} + d\sigma_{rr})((r + dr) \, d\theta \, dz)$$
$$- 2\sigma_{\theta\theta}\left(\frac{d\theta}{2}\right) dr \, dz], \tag{14.B.1}$$

where dz is the height of the volume element. The equation follows from the definition of the stress tensor as a force per unit area. From Newton's second law, this must be

$$F_r = \rho r \, dr \, d\theta \, dz \frac{d^2 u_r}{dt^2}, \tag{14.B.2}$$

where ρ is the density of the material. The equation of motion in the

r-direction is then

$$\rho \frac{d^2 u_r}{dt^2} = -\frac{\partial \sigma_{rr}}{\partial r} + \frac{\sigma_{\theta\theta} - \sigma_{rr}}{r}, \qquad (14.\text{B}.3)$$

which is just Newton's second law in cylindrical coordinates.

In order to proceed from this point, we must know something about the nature of the material in the arterial wall. This was discussed in Section 14.A. For the mathematical problems which follow, we shall make two assumptions about arterial walls. First, we shall assume that they are elastic, and hence obey Hooke's law. Second, we shall assume that they are composed of an incompressible material, so that

$$\nabla \cdot \mathbf{u} = 0.$$

We shall discuss these and other assumptions in more detail later. For the moment, we note simply that if they are true, Hooke's law takes the form (see Problem 12.5)

$$\sigma_{rr} = E \frac{\partial u_r}{\partial r},$$

$$\sigma_{\theta\theta} = E \frac{u_r}{r}, \qquad (14.\text{B}.4)$$

so that the equation of motion becomes

$$\rho \frac{\partial^2 u_r}{\partial t^2} = E \frac{\partial^2 u_r}{\partial r^2} + \frac{E}{r} \left[\frac{\partial u_r}{\partial r} - \frac{u_r}{r} \right], \qquad (14.\text{B}.5)$$

which can be solved for u_r.

Let us work out a couple of examples to see how the tube responds. For the first, let us take the simplest possible case—the case where the internal pressure is a constant. Then the left-hand side of Eq. (14.B.5) will vanish, and we will have a second-order equation for u_r. If R_i is the inner radius of the tube, and R_o the outer, then the two boundary conditions which must be satisfied are

$$\sigma_{rr} = -P \qquad (14.\text{B}.6)$$

at $r = R_i$, and

$$\sigma_{rr} = 0 \qquad (14.\text{B}.7)$$

at $r = R_o$. If we guess a solution for u_r of the form

$$u_r \propto r^l$$

and substitute back into Eq. (14.B.5), we find $l = \pm 1$, so that the most general solution for u_r is just

$$u_r = Ar + \frac{B}{r}. \tag{14.B.8}$$

If we now combine our definition of the stress tensor [Eq. (14.B.4)] with Eq. (14.B.7), we find

$$A = \frac{B}{R_o^2}, \tag{14.B.9}$$

while Eq. (14.B.6) yields

$$B = \frac{P}{E} \left(\frac{1}{R_o^2} - \frac{1}{R_i^2} \right)^{-1}, \tag{14.B.10}$$

so that the solution is

$$u_r = \frac{PR_i^2 R_o^2}{E(R_o^2 - R_i^2)} \left[\frac{r}{R_o^2} + \frac{1}{r} \right]. \tag{14.B.11}$$

There is an interesting sidelight to this result. Suppose we now ask for the value of the stress which is exerted axially around the tube at its outer boundary. This is just

$$\sigma_{\theta\theta}(r = R_o) = \frac{E}{R_o} u_r(r = R_o) = \frac{2P}{\left(\dfrac{R_o^2}{R_i^2} \right) - 1}. \tag{14.B.12}$$

Now if the tube is thin, so that

$$R_o = R_i + \delta,$$

we can expand the expression in the denominator and get

$$PR_i = \delta\sigma_{\theta\theta} = T_m, \tag{14.B.13}$$

where we have defined T_m as the *membrane tension*. This result should look very familiar. It is exactly the relation between the pressure and radius of curvature which was obtained in Chapter 5 for a fluid with surface tension. Thus, a very thin membrane can be thought of in the same way as surface tension—as a component of the system which tends to oppose increases in surface area.

It is more usual to be concerned with time-dependent pressure when dealing with blood flow. After all, the moving force behind the flow is the periodic pumping of the heart. How would the above analysis be changed if the pressure, instead of being constant, were time dependent? In this case, we would have to guess at a solution of Eq. (14.B.5) of the form

$$u_r = R(r)T(t), \tag{14.B.14}$$

where R is a function of r alone, and T a function of t. Inserting this into Eq. (14.B.5), we get

$$-\frac{\rho}{T}\frac{d^2T}{dt^2} = \frac{E}{R}\frac{d^2R}{dr^2} + \frac{E}{r}\left[\frac{dR}{dr} - \frac{1}{R}\right] = \lambda^2\rho, \tag{14.B.15}$$

where λ is a constant. The solution for T is then

$$T(t) = T_0 e^{i\lambda t}, \tag{14.B.16}$$

while the equation for R is just

$$\frac{d^2R}{dr^2} + \frac{1}{r}\frac{dR}{dr} - \frac{R}{r^2}\left(1 - \frac{\rho\lambda^2 r^2}{E}\right) = 0, \tag{14.B.17}$$

which is rather complicated. However, we can make a very reasonable approximation if we put in some numbers which are typical of blood flow. A typical value for λ, the frequency associated with the pressure, might be 7 rad/sec while values of the other parameters might be $\rho \sim 1.1$ g/cc, $R \sim 1$ cm, and $E \sim 10^6$ dynes/cm. Thus, we see that

$$\frac{\rho\lambda^2 R^2}{E} \ll 1,$$

so that the second term in parentheses in Eq. (14.B.17) can be dropped. In this case, Eq. (14.B.17) reduces to the equation for u_r which we had in the previous case, so that Eqs. (14.B.8) and (14.B.9) are again valid. Thus, the most general solution for u_r will be

$$u_r = \sum_\lambda a_\lambda\left(Ar + \frac{B}{r}\right)e^{i\lambda t}, \tag{14.B.18}$$

where we understand that we have taken an arbitrary sum of all possible solutions, and the summation is understood to extend over all allowed values of λ. These values will be determined by the boundary conditions.

Equation (14.B.6) still describes the boundary condition at the inner radius, but now the pressure is a function of t, and not a constant. In general, the pressure will be some time-dependent function. If we Fourier analyze it, we can write

$$P(t) = \sum P_n e^{i\omega_n t},$$

where $\omega_n = 2\pi n/\tau$, and τ is the period of the pulse.

If we now impose the boundary condition at the inner surface, we have

$$B\left(\frac{1}{R_o^2} - \frac{1}{R_i^2}\right) \sum_\lambda a_\lambda e^{i\lambda t} = -\frac{1}{E} \sum_n P_n e^{i\omega_n t}. \qquad (14.B.19)$$

In order to satisfy Eq. (14.B.19) for any value of t, we must have

$$B\left(\frac{1}{R_o^2} - \frac{1}{R_i^2}\right) a_\lambda = -\frac{P_n}{E}, \qquad (14.B.20)$$

$$\lambda = \omega_n,$$

so that, for our final solution, we have

$$u_r = \frac{R_i^2 R_o^2 P(t)}{E(R_o^2 - R_i^2)} \left[\frac{r}{R_o^2} + \frac{1}{r}\right], \qquad (14.B.21)$$

which is identical to Eq. (14.B.11), except that now the pressure is understood to be time dependent.

One interesting result can be seen immediately from this equation. When we are dealing with physiological systems, we often cannot measure quantities of direct interest, but must infer them from indirect measurements. For example, it is often not convenient to measure pressure inside of an artery directly (although this can be easily done) and one might wish to know the pressure just from observing the outer wall of the artery. How will it move as the pressure is applied? If we define $u_r(r = R_o)$ as the distance the outer wall will move, then

$$\frac{u_r(r = R_o)}{R_o} = \frac{2P(t)}{E} \left[\left(\frac{R_o}{R_i}\right)^2 - 1\right]^{-1}.$$

In other words, for a perfectly elastic artery wall, the outer surface will move *in phase* with the internal pressure. For a viscoelastic material, however, this will not be the case, since there will be a time lag while the wall responds to the changes in pressure which will, in turn, be reflected by a phase lag.

With this introduction to the behavior of arterial walls, we will turn to the problem of describing the flow of blood in an artery.

C. BLOOD FLOW IN AN ARTERY

We are now in a position to write down the equations which govern the flow of blood in an artery. For the equations which describe the blood itself, of course, we have the Navier–Stokes equation

$$\frac{\partial \mathbf{v}}{\partial t} + (\mathbf{v} \cdot \nabla)\mathbf{v} = -\frac{1}{\rho}\nabla P + \frac{\eta}{\rho}\nabla^2 \mathbf{v} \qquad (14.C.1)$$

and the equation of continuity

$$\frac{\partial \rho}{\partial t} + \nabla \cdot (\rho \mathbf{v}) = 0. \tag{14.C.2}$$

For the arterial walls, the equations are simply the obvious generalizations of Eq. (14.B.3), so that

$$\rho_\omega \frac{\partial^2 u_r}{\partial t^2} = \frac{\partial \sigma_{rr}}{\partial r} + \frac{\partial \sigma_{rz}}{\partial z} + \frac{\sigma_{rr} - \sigma_{\theta\theta}}{r} \tag{14.C.3}$$

and

$$\rho_\omega \frac{\partial^2 u_z}{\partial t^2} = \frac{\partial \sigma_{rz}}{\partial r} + \frac{\partial \sigma_{zz}}{\partial z} + \frac{\sigma_{rz}}{r}, \tag{14.C.4}$$

where ρ_ω is the density of the materials in the wall. ρ_ω will, in general, satisfy a continuity equation like Eq. (14.C.2).

These equations must be solved subject to boundary conditions which we shall discuss in detail later. As they stand, they are extremely difficult to solve. They are badly nonlinear, and in Eq. (14.C.1), the coefficient η is, in general, a function of the velocity. Nevertheless, if we want to find a simple solution to the problem, we will have to make some approximations. The first of these will be to assume that blood is a classical incompressible Newtonian fluid, so that

$$\eta = \text{const.} \tag{14.C.5}$$

and

$$\nabla \cdot \mathbf{v} = 0. \tag{14.C.6}$$

Second, we will assume that the nonlinear term in the Navier–Stokes equation can be dropped. This corresponds to assuming that the viscous terms are quite large, so that

$$(\mathbf{v} \cdot \nabla)\mathbf{v} \ll \frac{\eta}{\rho} \nabla^2 \mathbf{v}. \tag{14.C.7}$$

Finally, throughout the section, we will assume that there is complete azimuthal symmetry, so that nothing depends on the coordinate angle. The equations of motion for the fluid then become (see Problem 14.1)

$$\rho \frac{\partial v_r}{\partial t} = -\frac{\partial P}{\partial r} + \frac{\eta}{\rho} \frac{\partial}{\partial r} \left[\frac{1}{r} \frac{\partial}{\partial r}(r v_r) + \frac{\partial^2 v_r}{\partial z^2} \right]$$

$$\rho \frac{\partial v_z}{\partial t} = -\frac{\partial P}{\partial z} + \frac{\eta}{\rho} \left[\frac{1}{r} \frac{\partial}{\partial r} \left(r \frac{\partial v_z}{\partial r} \right) + \frac{\partial^2 v_z}{\partial z^2} \right] \tag{14.C.8}$$

$$\frac{\partial v_z}{\partial z} + \frac{1}{r} \frac{\partial}{\partial r}(r v_r) = 0.$$

This leads us to an equation for the pressure (see Problem 14.2) which is of the form

$$\nabla^2 P = \frac{1}{r}\frac{\partial}{\partial r}\left(r\frac{\partial P}{\partial r}\right) + \frac{\partial^2 P}{\partial z^2} = 0. \tag{14.C.9}$$

Following the standard procedure outlined in Eq. (14.B.14), we assume that the solution is of a separable form

$$P = R(r)Z(z), \tag{14.C.10}$$

and find that

$$-k^2 = \frac{1}{Z}\frac{d^2 Z}{dz^2} = -\frac{1}{R}\cdot r\frac{\partial}{\partial r}\left(r\frac{\partial R}{\partial r}\right), \tag{14.C.11}$$

where k is an arbitrary constant. We then see immediately that

$$Z(z) = e^{ikz}, \tag{14.C.12}$$

and are left with an equation for R of the form

$$\frac{d^2 R}{dr^2} + \frac{1}{r}\frac{dR}{dr} - k^2 R = 0, \tag{14.C.13}$$

which is just Bessel's equation of order zero (this can be seen by writing $-k^2 = (ik)^2$).

Thus, the pressure is given by

$$P(r, z, t) = A J_0(ikr) e^{ikz} e^{i\omega t}, \tag{14.C.14}$$

where A is an arbitrary constant, and we have followed the procedure outlined in Section 14.B and assumed that all time dependences are of the form $e^{i\omega t}$.

We can then put this solution for P back into Eq. (14.C.8) and solve for v_z. If we assume a form for v_z such as

$$v_z = v_z(r) e^{ikz} e^{i\omega t}, \tag{14.C.15}$$

then we find the equation governing $v_z(r)$ to be

$$\frac{d^2 v_z}{dr^2} + \frac{1}{r}\frac{dv_z}{dr} - \gamma^2 v_z = -\frac{ikA}{\eta} J_0(ikr), \tag{14.C.16}$$

where we have defined

$$\gamma^2 = k^2 + \frac{i\omega\rho}{\eta}. \tag{14.C.17}$$

Now Eq. (14.C.16) is of the familiar form whose general solution is the sum of a particular and a homogeneous solution. The homogeneous solution, in analogy to the solution to Eq. (14.C.9) is simply

$$v_z^h = J_0(i\gamma r).$$

If we guess that the form of the particular solution is

$$v_z^p = BJ_0(i\kappa r),$$

then B can be determined by plugging back into Eq. (14.C.8) to be

$$B = \frac{k}{\omega\rho} A,$$

so that the most general form for v_z is just

$$v_z(r, z, t) = \left[\frac{kA}{\omega\rho} J_0(ikr) + CJ_0(i\gamma r)\right] e^{ikz} e^{i\omega t}. \tag{14.C.18}$$

It must be emphasized that in this expression, only C is unknown. The constant A will be determined by the boundary conditions on the pressure at $z = 0$.

Following similar steps, it is shown in Problem 14.3 that the radial velocity is just

$$v_r(r, z, t) = \left[\frac{kA}{\omega\rho} J_1(ikr) + DJ_1(i\gamma r)\right] e^{i\omega t} e^{ikz}. \tag{14.C.19}$$

Thus, we have solved Eqs. (14.C.1) and (14.C.2) which describe the motion of the fluid in the artery, subject to the approximations in Eqs. (14.C.5), (14.C.6), and (14.C.7). Let us examine the equations of motion for the walls before we start discussing the boundary conditions. As we did for the fluid, we will make some approximations to simplify our work. We shall assume that we are dealing with a purely elastic solid which obeys Hooke's law, and that the solid is incompressible, so that

$$\nabla \cdot \mathbf{u} = 0. \tag{14.C.20}$$

In this case, components of the stress tensor which we need are given by

$$\sigma_{rr} = E\frac{\partial u_r}{\partial r},$$

$$\sigma_{\theta\theta} = E\frac{u_r}{r}, \tag{14.C.21}$$

$$\sigma_{rz} = \frac{1}{2}E\left[\frac{\partial u_r}{\partial z} + \frac{\partial u_z}{\partial r}\right],$$

so that the equations of motion are

$$\rho_\omega \frac{\partial^2 u_r}{\partial t^2} = E \frac{\partial^2 u_r}{\partial r^2} + 2E\left[\frac{\partial^2 u_r}{\partial z^2} + \frac{\partial^2 u_z}{\partial r \partial z}\right],$$
$$+ \frac{E}{r}\frac{\partial u_r}{\partial r} - \frac{E}{r^2}u_r, \tag{14.C.22}$$

$$\rho_\omega \frac{\partial^2 u_z}{\partial t^2} = 2E\left[\frac{\partial^2 u_r}{\partial r^2} + \frac{\partial^2 u_z}{\partial r \partial z}\right] + E\frac{\partial^2 u_z}{\partial z^2}$$
$$+ \frac{E}{2r}\left[\frac{\partial u_r}{\partial z} + \frac{\partial u_z}{\partial r}\right]. \tag{14.C.23}$$

These equations are still very complicated because they are coupled. Although they could be solved numerically, we will look for further approximations which might give us an easy solution.

We know from Section 14.B that when a pulse moves down the artery, the artery will expand and contract. It seems reasonable to assume that most of this motion is in the radial direction, and that the stretching of the artery in the z-direction is probably less pronounced. We will take this physical idea to its extreme, and assume that

$$u_z \ll u_r. \tag{14.C.24}$$

It must be noted that this approximation, while it does simplify the equations of motion, does a certain amount of violence to our initial assumptions, since it assumes that there is an anisotropy in the arterial wall which restricts motion in the z-direction. Because of this, we will drop considerations of u_z from this point on.

With Eq. (14.C.24), the equation of motion becomes

$$\frac{\partial^2 u_r}{\partial r^2} + \frac{1}{r}\frac{\partial u_r}{\partial r} + \left[\frac{\omega^2 \rho_\omega}{E} - 2Ek^2 - \frac{1}{r^2}\right]u_r = 0, \tag{14.C.25}$$

which has the solution

$$u_r(r, z, t) = FJ_1(i\Gamma r)e^{ikz}e^{i\omega t}, \tag{14.C.26}$$

where we have defined

$$\Gamma^2 = 2Ek^2 - \frac{\omega^2 \rho_\omega}{E}. \tag{14.C.27}$$

With these solutions, we must turn to the boundary conditions. We will start with those which must be imposed on the fluid. We know from symmetry that at $r = 0$ we must have

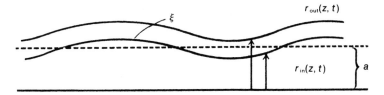

Fig. 14.4. The arterial wall during pulsatile movement.

$$v_r(r = 0) = 0,$$

$$\frac{\partial v_z}{\partial r}(r = 0) = 0,$$

(14.C.28)

since the fluid may not flow away from the center. These conditions are automatically satisfied by Eqs. (14.C.18) and (14.C.19). The other boundary conditions concern the inner surface of the artery (see Fig. 14.4).

The general boundary condition here is that the relative motion between the fluid and the wall must vanish at this surface in keeping with our ideas about the nature of viscosity. Thus, at $r = r_{in}$, we must have

$$v_r = \frac{\partial u_r}{\partial t},$$

(14.C.29a)

$$v_z = \frac{\partial u_z}{\partial t}.$$

(14.C.29b)

At this surface, we must also have a situation where the stresses are continuous. The reader may compare this to the boundary condition which was imposed in the derivation of Love waves in Section 12.F. The sheer stress along a surface on the inner face of the artery exerted by the wall is just

$$\sigma_{rz} = E\left(\frac{\partial u_r}{\partial z} + \frac{\partial u_z}{\partial r}\right)_{r=r_{in}}$$

(14.C.30)

while in Problem 14.4, we show that the stress exerted by the fluid is just

$$\sigma'_{rz} = \eta\left(\frac{\partial v_r}{\partial z} + \frac{\partial v_z}{\partial r}\right),$$

(14.C.31)

so that we must have

$$E\left(\frac{\partial u_r}{\partial z} + \frac{\partial u_z}{\partial r}\right) = \eta\left(\frac{\partial v_r}{\partial z} + \frac{\partial v_z}{\partial r}\right),$$

(14.C.32)

at $r = r_{in}$. A similar argument for stresses in the radial direction yields the result that

$$-P + 2\eta \frac{\partial v_r}{\partial r} = 2E \frac{\partial u_r}{\partial r}. \tag{14.C.33}$$

At the outer boundary, the stresses exerted by the arterial wall must be continuous with those exerted by the surrounding medium, and must vanish if that medium is a vacuum (see Problem 14.5).

We now come to the third important set of complications in our theory. In general, the deformation of the artery is not small compared to its radius, so the full time dependence of r_{in} and r_{out} must be included in writing down the boundary conditions. In addition, the general shape of arteries need not be cylinders of constant cross section. In fact, the aorta, the artery leading away from the heart, has a general shape like that shown in Fig. 14.5. Therefore, in order to apply the boundary conditions easily, we shall assume that for the artery in question,

$$r_{in} = a + \xi,$$

where a is constant, and that the deformations are small, so that

$$\frac{\xi}{a} \ll 1. \tag{14.C.34}$$

We are now in a position to apply the boundary conditions. In general, we wish to treat the constant A which appears in Eq. (14.C.14) as given by the initial conditions (as we did in Section 14.B), so that we want to determine the constants C, D, and F from Eqs. (14.C.18), (14.C.19), and (14.C.26).

From the condition in Eq. (14.C.29a), together with our approximation of ignoring u_z, we have

$$v_z(r = a) = 0, \tag{14.C.35}$$

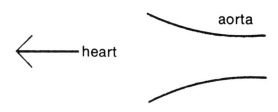

Fig. 14.5.   A typical shape of an aorta.

which gives

$$C = -\frac{k}{\omega\rho} A \frac{J_0(ika)}{J_0(i\gamma a)},$$ (14.C.36)

and determines one of the three constants.

Equation (14.C.29b) gives

$$\frac{k}{\omega\rho} AJ_1(ika) + DJ_1(i\gamma a) = i\omega FJ_1(\Gamma a),$$ (14.C.37)

which gives one relation between the other two constants. The third relation comes from the stress condition in Eq. (14.C.33).

$$A\left[-J_0(ika) + \frac{2ik^2\eta}{\omega\rho} J_1'(ika) \right] + i\gamma DJ_1'(i\gamma a) = 2iEF\Gamma J_1'(\Gamma a)$$ (14.C.38)

where the prime denotes differentiation of the Bessel function with respect to its argument.

Thus, by making a large number of approximations, we were able to come to a solution of the general problem. It should be obvious to the reader that none of these approximations stands on very firm physical grounds, so that in realistic work, they would have to be examined very carefully. However, it is hoped that working through the simplest possible case of arterial flow has demonstrated the techniques which must be employed to link up the studies of fluid mechanics which were treated in the first part of the text and the studies of elastic solids which were discussed in the second. The working out of an actual example is left to Problem 14.6.

The problem of the flow in the circulatory system is receiving a great deal of attention in current research. Most of the work that is done is logically quite similar to what has been discussed in this section. The general technique is to make as many approximations as one can, trying always to treat the quantities of interest as exactly as possible. We mention a few examples to illustrate this point.

(1) *Arteriosclerosis.* This is the problem of the flow of blood through an artery which can be partially obstructed by deposits. In this case, the assumption that arterial walls were almost circular cylinders of constant cross sections [Eq. (14.C.34)] would no longer be useful, since we wish to study the effect of changes in arterial cross section on the flow. What is usually done in this case is to assume that the walls are rigid, but of a deformed shape, and then try to proceed as realistically as possible.

(2) *The Entry Problem.* This is the problem which is concerned with the way the velocity profile develops from the point at which the blood enters (e.g., at the heart) until it is fully developed. In this case, the procedure of dropping the nonlinear terms [Eq. (14.C.6)] will not be useful, since we are trying to examine changes in the velocity itself. In this problem, one usually keeps the nonlinear terms, and keeps the approximation that the arterial walls are rigid and of uniform cross section.

The point of these examples is that even though the general problem of flow in the circulatory system is too complicated to solve with present techniques, a great deal of progress can be made in isolating individual aspects of the problem and solving them. In each case, the simplifying assumptions have to be chosen in such a way as to retain a realistic description of the phenomenon we are trying to describe, and treat other aspects of the problem as realistically as possible.

It should be reassuring to the student that even though the science of hydrodynamics was developed over a century ago, there are still important problems waiting to be solved.

D. THE URINARY DROP SPECTROMETER

Another, more speculative application of the techniques which we have learned in this text to an area of medicine is the urinary drop spectrometer. This is an instrument whose function is to provide early diagnoses of abnormalities in the urinary tract.

In Fig. 14.6 is presented a simplified sketch of the lower urinary tract. The urine from the kidneys is stored in the bladder, and passes to the

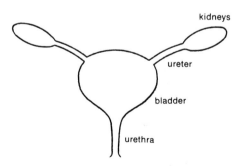

Fig. 14.6. A schematic diagram of a urinary system.

outside through a deformable tube called the urethra. Since the urethra is open to the outside, it is constantly being invaded by bacteria. Urination performs the important function of washing these bacteria out.

Clearly, obstructions or impediments to the flow will give the bacteria a chance to cause infections in the urethra, which will, in turn, weaken the tissue and make the system more susceptible to infection at a later date. Over the course of years, these infections can progress to the bladder, and even the kidneys. For this reason (as well as for many others which are equally compelling), it is important to be able to develop a diagnostic technique for detecting these small obstructions and impediments *before* they have a chance to cause a great deal of damage.

The urinary drop spectrometer is such a technique. It works on the following principle: The stream of urine passes through the urethra during the process of urination, and flows around the obstruction. Information about the obstruction is then contained in the stream, which emerges and breaks into drops. It is a reasonable assumption that some of the information about the obstruction is transmitted to these drops. If we then arrange things so that the drops interrupt a light beam between a light source and a phototube (see Fig. 14.7), then each drop will correspond to a pulse in the output of the tube. If we knew how to analyze this set of pulses, we would be able to gather information about the condition of the urethra from a normal urination. Such a technique, if it were perfected, would be something like a chest X-ray for the urinary system—it could be a routine part of a physical examination, and could give early warning of urinary tract difficulties.

Obviously, the hydrodynamic problems associated with the transfer of information about the obstruction to the drops are extremely difficult. The flow in elastic tubes was considered in the previous section. Once the stream emerges from the urethra, however, an entirely new set of considerations comes into play. We then have a cylindrical tube of fluid moving along under the influence of two forces: The pressure of the fluid and the surface tension. Such a system is called a *capillary jet*. In the next

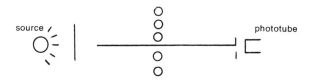

Fig. 14.7.   The urinary drop spectrometer.

section, we will look at the simplest such jet—one in which the fluid moves everywhere with constant velocity—and try to understand why it breaks into drops. The question of how the drops are formed, and how they can be related to urethral obstructions, is not understood at the present time.

E. STABILITY OF A CAPILLARY JET

Consider a jet of incompressible fluid of density ρ and surface tension T moving with constant velocity c to the right (see Fig. 14.8). Without loss of generality, we can take $c = 0$, since a simple Galilean transformation will change c to any value we choose. Let us further assume that the fluid is inviscid ($\eta = 0$) for the sake of simplicity, and that the cross section of the unperturbed jet is a circle of radius a.

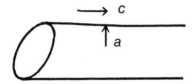

Fig. 14.8. The unperturbed jet.

The equilibrium for such a jet is clearly one in which $v_z = v_r = 0$, and the pressure is a constant given by

$$P = \frac{T}{a}. \tag{14.E.1}$$

To examine the question of stability, we will use the technique of Section 4.D and introduce small, time-dependent perturbations to the system, and see under what conditions they might be expected to grow. We will also assume that the perturbations which we introduce are irrotational, so that we can write

$$\mathbf{v} = \nabla\phi, \tag{14.E.2}$$

where ϕ is the velocity potential. As in Section 4.B, the equation for the velocity potential is just

$$\nabla^2\phi = 0. \tag{14.E.3}$$

It should be noted that the velocities referred to in the above equations are the perturbing velocities, since the equilibrium velocities are zero. Let us look at perturbations of the form

$$\phi(r, \theta, z, t) = \phi_1(r, \theta) \cos kz \cos \sigma t, \qquad (14.\text{E}.4)$$

with the understanding that we can, if we wish, regard this as one component of a Fourier series expansion of any actual perturbation. The condition that the jet be stable is that

$$\sigma^2 \geq 0, \qquad (14.\text{E}.5)$$

since in this case, there will be no growth of the perturbation with time.

The other equations which we have at our disposal are the Euler equation in the form

$$\frac{P}{\rho} = \frac{\partial \phi}{\partial t} + \text{const.}, \qquad (14.\text{E}.6)$$

where we have dropped second-order terms in the velocity, and the condition at the surface which states that

$$P = T\left(\frac{1}{R_1} + \frac{1}{R_2}\right), \qquad (14.\text{E}.7)$$

where R_1 and R_2 are the principle radii of curvature.

If we insert our assumed form of the perturbation into Eq. (14.E.3), we find

$$\frac{1}{r^2}\frac{\partial^2 \phi_1}{\partial r^2} + \frac{1}{r}\frac{\partial \phi_1}{\partial r} + \frac{1}{r^2}\frac{\partial^2 \phi_1}{\partial \theta^2} - k^2 \phi_1 = 0, \qquad (14.\text{E}.8)$$

so that, if we assume a separable form of the solution

$$\phi_1 = k(r)\Theta(\theta),$$

and proceed as usual, we find that

$$\frac{\partial^2 \Theta}{\partial \theta^2} = -s^2, \qquad (14.\text{E}.9)$$

which means that

$$\Theta \propto \cos s\theta,$$

where s is the separation constant analogous to the constant k in Eq. (14.C.11). The equation for $R(r)$ is then

$$\frac{d^2 R}{dr^2} + \frac{1}{r}\frac{dR}{dr} + \left[(ik)^2 - \frac{s^2}{r^2}\right] R = 0, \qquad (14.\text{E}.10)$$

which is just the Bessel equation [compare with Eq. (14.C.13)]. The solutions will be (assuming that the velocity potential remains finite at

$r = 0$)

$$R \propto I_s(kr),$$

so that the final expression for the velocity potential is

$$\phi = AI_s(kr) \cos s\theta \cos kz \cos \sigma t. \qquad (14.E.11)$$

In these expressions, the function $I_s(kr)$ is called the modified Bessel function, and is identical to $e^{i\pi s/2} J_s(ikr)$.

This result, together with Eqs. (14.E.2) and (14.E.6), completely defines all of the hydrodynamic variables in the problem up to a constant. To proceed further, it is necessary to apply the boundary conditions.

When the perturbations are applied, the surface of the jet will be deformed from a perfect circular cylinder (see Fig. 14.9). If we write

$$r = a + \xi, \qquad (14.E.12)$$

then ξ is a small parameter representing this deviation. From the conditions that a fluid element in the surface moves with the same velocity as the surface itself, we have

$$\frac{\partial \xi}{\partial t} = \frac{\partial \phi}{\partial r} \qquad (14.E.13)$$

at $r = a$, while from Problem 14.7 and Eq. (14.E.7), we have the condition that

$$P(r = a) = T\left[\frac{1}{a} - \frac{1}{a^2}\left(\xi + \frac{\partial^2 \xi}{\partial \theta^2}\right) - \frac{\partial^2 \xi}{\partial z^2}\right]. \qquad (14.E.14)$$

From Eq. (14.E.13), we have the result that

$$\xi = \frac{kA}{\sigma} I_s'(ka) \cos s\theta \cos kz \sin \sigma t, \qquad (14.E.15)$$

while Eq. (14.E.6) gives

$$P = \frac{A}{\sigma} I_s(kr) \cos s\theta \cos kz \cos \sigma t. \qquad (14.E.16)$$

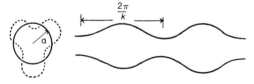

Fig. 14.9. End and side views of the perturbed jet.

If we now combine Eqs. (14.E.14), (14.E.15), and (14.E.16), we get

$$\sigma^2 = ka\frac{I_s'(ka)}{I_s(ka)}[k^2a^2 + s^2 - 1]\frac{T}{\rho a^3}. \qquad (14.E.17)$$

It is this equation which determines the time dependence of the perturbation, and hence the stability of the system.

It is a property of the Bessel functions that

$$\frac{I_s'(ka)}{I_s(ka)} > 0$$

for any value of the argument. Hence the jet can be unstable only if

$$k^2a^2 + s^2 - 1 < 0, \qquad (14.E.18)$$

in which case

$$\cos \sigma t = \frac{e^{i\sigma t} + e^{-i\sigma t}}{2} \rightarrow e^{|Im\,\sigma|t}.$$

From Eq. (14.E.9), it is clear that the constant s must be an integer. Otherwise, the solution for Θ would not be single valued. This means that if s has *any* nonzero value, Eq. (14.E.18) can never be satisfied, and the perturbation will not grow in time. Thus, perturbations like that in Figs. 14.10(a) and 14.10(b), in which the jet is "fluted," either with or without a z-dependence, will not grow with time, but will simply oscillate around equilibrium.

Fig. 14.10(a). A "fluted" perturbation of the jet.

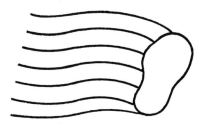

Fig. 14.10(b). A "fluted" perturbation with a z-dependence.

If $s = 0$, however, so that we consider only axially symmetric perturbations, then Eq. (14.E.18) can be satisfied provided that

$$ka < 1. \tag{14.E.19}$$

Since $k = 2\pi/\lambda$, this means that the perturbations whose wavelength satisfies the condition $ka < 1$ will grow exponentially with time. This is known as the *Rayleigh criterion* for jet stability.

Thus, we have shown that the capillary jet is indeed unstable, and will break up into drops at some time. We have also shown that the perturbations to which the jet is unstable are those which are axially symmetric and whose wavelength is longer than the circumference of the unperturbed jet.

Equation (14.E.17) tells us how fast each perturbation grows with time. Since $\sigma = 0$ at $ka = 0$ and $ka = 1$, there must be a maximum value which σ can attain. Numerical analysis shows that this occurs when

$$\lambda \approx 9.2a.$$

A first guess at the drops which would be formed, then, would be to assume that this fastest growing perturbation outstrips all of the others, and that the breakup process is dominated by this single wavelength perturbation at large times. In this case, we would expect equally spaced drops of equal mass when the jet finally disintegrates. Deviations from this expectation would presumably be due to the presence of other effects, among which might be the obstruction in the urethra through which the fluid has passed.

PROBLEMS

14.1. Show that in the case of cylindrical symmetry, the Navier–Stokes equation can be written as in Eq. (14.C.8).

14.2. Show that the equation for the pressure given in Eq. (14.C.9) follows from the Navier–Stokes equation, continuity and the approximations discussed in the text.

14.3. Derive Eq. (14.C.9) for the radial velocity of the fluid.

14.4. From the definition of the tensor σ'_{ik} in Chapter 8, show that the axial stress exerted by the fluid at the inner radius is

$$\eta \left(\frac{\partial v_r}{\partial z} + \frac{\partial v_z}{\partial r} \right),$$

and that the radial stress is

$$- P + 2\eta \frac{\partial v_r}{\partial r}.$$

14.5. Calculate the radial stress exerted by the artery in Section 14.C at the outer radius of the vessel. This must be continuous with the stress exerted by the surrounding medium. Is it possible for the surrounding medium to be a vacuum? What does this tell you about the assumption that u_z can be neglected in that case?

14.6. Consider the artery in Section 14.C in the case when the wall is rigid. This is the limit $F = 0$, $E \to \infty$. Calculate the total flow, given by

$$Q = 2\pi \int v_z r \, dr$$

in the limit of steady flow, given by

$$ka \ll \sqrt{\frac{\omega \rho a}{\eta}} \ll 1.$$

(*Hint*: The limiting form of the Bessel function for small argument is

$$J_0(x) \approx 1 - \frac{x^2}{4} + \cdots,$$

and compare it to the Poisieulle result.)

14.7. Consider a deformed cylinder, as shown in Fig. 14.11. Let R be the radius of curvature at a point, and r the distance from the center to that point. Let Δs be the arc length along the actual surface (shown as a solid line), and $r\Delta\theta$ the arc length along the surface shown as a dotted line.

(a) Show that

$$\frac{\Delta s}{\Delta \theta} = r\left[1 + \left(\frac{1}{r}\frac{\Delta r}{\Delta \theta}\right)^2\right]^{1/2}.$$

(b) Hence show that for small deformations,

$$R = r\left(1 + \frac{1}{r}\frac{\partial^2 r}{\partial \theta^2}\right).$$

(c) Hence derive Eq. (14.E.14), given Eq. (14.E.12).

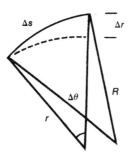

Fig. 14.11.

14.8. Let us see if we can come to a simple understanding of the Rayleigh condition for jet instability in Eq. (14.E.14). Consider a film whose surface tension is T, and which is deformed in an axially symmetric way as in Fig. 14.12. Let the equation of the surface be given by

$$r = a + b \cos kz,$$

where $k = 2\pi/\lambda$.

(a) Show that at the point A, where the surface is maximally deformed outward, the pressure is

$$T\left[\frac{1}{a+b} + \frac{4\pi^2 b}{\lambda^2}\right] = P_A.$$

(b) Show that at B, where the surface is maximally deformed inward, the pressure is

$$T\left[\frac{1}{a-b} - \frac{4\pi^2 b}{\lambda^2}\right] = P_B.$$

(c) Hence show that if $b \ll a$, the film will be unstable unless $2\pi a > \lambda$.

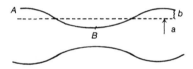

Fig. 14.12.

14.9. Calculate the Reynolds number for typical blood flow in a human artery, and for typical flow in the urethra.

14.10. One of the problems discussed in connection with the urinary drop spectrometer is the question of whether, in passing through the air, the urine stream picks up a static charge. Calculate the effect of a static surface charge density σ on the Rayleigh equation (14.E.7).

14.11. The development of the Rayleigh theory assumed that the jet existed in a vacuum. This, of course, is not the case.

(a) Assuming that the jet is proceeding through a stationary atmosphere of density ρ', find the atmospheric pressure at the surface of the distorted jet.

(b) Hence modify Eq. (14.E.14) to take account of aerodynamic effects.

(c) How is the Rayleigh equation changed by the inclusion of this effect?

14.12. Let us consider the stability of blood flow in an artery. Suppose that in equilibrium flow, the velocity is entirely in the z-direction, and is given by a function $U(r)$. Let us then consider a small perturbation whose stream function is of the form

$$\psi(r, z, t) = \phi(r) e^{i(kz - \beta t)}.$$

(a) Find the small perturbation velocities v_z and v_r.

(b) Show that we can define a stream function for this problem provided that there is azimuthal symmetry.

(c) Show that the equation for ϕ is

$$(U - c)(\phi'' - k^2\phi) - U''\phi = -\frac{i}{kR}(\phi'''' - 2k^2\phi' + k^4\phi)$$

where we have defined $c = \beta/k$ and R is the Reynolds number. This is called the Orr–Sommerfeld equation, and is widely used in studying stability.

(d) What are the boundary values for ϕ?

14.13. Show that if we neglect terms of order $1/R$, the Orr–Sommerfeld equation will have a singular point when $U = c$. If r_k is the distance to the point where this occurs, show that the perturbation velocity in the z-direction must go as

$$u_z \sim \ln(r - r_k)$$

near $r = r_k$. Can you give a reason why this singularity occurs, and how it can be removed? (*Hint*: Remember the discussion connected with boundary layers.) The problem of the transition from laminar to turbulent flow is dealt with in great detail in the text by Schlichting included in the references.

14.14. The problem of flow through a constricted tube is, in general, a very difficult one. Suppose that the radius of an artery is given by

$$R(z) = R_0 - \delta f\left(\frac{z}{z_0}\right),$$

where the function f defines the narrowing of the artery in some region.

(a) Show that if $\delta/z_0 \ll 1$ and $\delta/R_0 \ll 1$, the Navier–Stokes equations reduce to

$$v_z \frac{\partial v_z}{\partial z} + v_z \frac{\partial v_r}{\partial r} = -\frac{1}{\rho}\frac{\partial P}{\partial z} + \nu\left(\frac{\partial^2 v_r}{\partial r^2} + \frac{1}{r}\frac{\partial v_r}{\partial r}\right)$$

and

$$\frac{\partial P}{\partial r} = 0.$$

(b) Hence show that

$$\frac{d}{dz}\int_0^R rv_r^2\, dr = -\frac{1}{\rho}\frac{dP}{dz}\frac{R^2}{2} + \nu R\left(\frac{\partial v_r}{\partial r}\right)_R.$$

(c) If we define

$$\eta = \frac{R - r}{R}$$

and

$$v_r(r = 0) = u,$$

and assume a form for the solution

$$\frac{v_r}{u} = A\eta + B\eta^2 + C\eta^3 + D\eta^4 + E,$$

show that applying the boundary conditions to determine the constants yields

$$\frac{v_r}{u} = \left(-\frac{\lambda + 10}{7}\right)\eta + \left(\frac{3\lambda + 5}{7}\right)\eta^2 + \left(-\frac{3\lambda - 12}{7}\right)\eta + \left(\frac{\lambda + 4}{7}\right)\eta^4,$$

where

$$\lambda = \frac{R^2 \rho}{\nu u}\frac{dP}{dz},$$

and dp/dz is determined by demanding consistency with the equation in part (b).

REFERENCES

S. Middleman, *Transport Phenomena in the Cardio-vascular System*, Wiley Interscience, New York, 1972.
> A good description of the physical processes involved in circulation. The text suffers somewhat from a rather inelegant use of mathematics.

H. Schlichting, (*op. cit.*—see Chapter 8).
> Contains some useful presentations of flow in tubes and stability criterion.

For a discussion of the capillary jet, see

H. Lamb, (*op. cit.*—see Chapter 1).

N. Bohr, *Phil. Trans. Roy. Soc., London* **A209**, 281 (1909).

For a discussion of Bessel functions, see

G. N. Watson, *A Treatise on the Theory of Bessel Functions*, Cambridge U.P., 1958.
> Probably the most complete work of this type in existence. Had it been written later, it could have been titled "Everything You Have Always Wanted to Know About Bessel Functions."

M. Abramowitz and I. A. Stegun (eds.), *Handbook of Mathematical Functions*, U.S. Department of Commerce.
> Chapters 9 and 10 give a complete (but concise) summary of the properties of Bessel and related functions. This is one of the best and most useful reference books on mathematical functions.

J. D. Jackson, *Classical Electrodynamics*, John Wiley and Sons, New York, 1962.
> In Chapter 3 there is a good presentation of Bessel's equation and its solutions in the context of a physical problem.

For a discussion of the Urinary Drop Spectrometer, see

G. Aiello, P. La France, R-C. Ritter, and J. S. Trefil, *Physics Today*, September 1974.

Appendices

Merely corroborative details to lend an aspect of
verisimilitude to what would otherwise be a bald
and unconvincing narrative.

GILBERT AND SULLIVAN
The Mikado

INTRODUCTION

Throughout most of the text, an effort has been made to make the
mathematical development of the various topics self-contained. Inevita-
bly, however, there will be students who, for one reason or another, have
missed some of the mathematical background necessary for the discussions. The purpose of these appendices is to provide a quick reference in
the mathematics which is used throughout the text, particularly for
Cartesian tensor notation, differential equations, and expansions in series.
These appendices do not constitute a textbook in mathematical physics,
however, and are included primarily because in teaching this material I
have found that some students can benefit from a short presentation of
the main mathematical techniques. Students wishing more detail, or
wishing to pursue these topics further, are referred to the following
standard texts:

Jon Mathews and R. L. Walker, *Mathematical Methods of Physics*, W. A. Benjamin, New
 York, 1970. A readable and concise treatment of mathematical physics, which should be
 easy for the student to follow.
W. Morse and H. Feschbach, *Methods of Theoretical Physics*, McGraw-Hill, New York,
 1953. A two volume treatise which contains almost anything the average physicist needs
 in the way of mathematical techniques.

APPENDIX A CARTESIAN TENSOR NOTATION

Throughout the text, it is frequently found useful to use tensor rather than vector notation. In this appendix, this type of notation will be explained.

A vector is usually considered to be a quantity which has both magnitude and direction, and can be specified by giving its length and the angles defining its directions (see Fig. A.1). However, a vector can also be completely specified by listing its three components (for our purposes, we take these to be the x-, y-, and z-components. Thus, we could write

$$\mathbf{V} = (V_x, V_y, V_z)$$

or, more simply,

$$\mathbf{V} = V_i,$$

where the index i is understood to run from 1 to 3, where V_1 is the x-component of the vector, V_2 the y-component, and V_3 the z-component. This is the simplest example of Cartesian tensor notation.

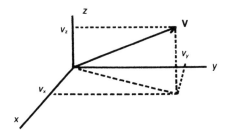

Fig. A.1.   A vector in three dimensions.

Now it is well known that if the coordinate system is rotated, the components of the vector in the new system (which we will call $\mathbf{V}'$) are related to the coordinates in the old system by the relation

$$\mathbf{V}' = \mathbf{R}\mathbf{V}, \tag{A.1}$$

where $\mathbf{R}$ is the matrix which describes the coordinate transformation. Recalling the definition of matrix multiplication, the component V'_i is just

$$V'_i = \sum_{j=1}^{3} R_{ij} V_j. \tag{A.2}$$

For a rotation in two dimensions, for example, a rotation through an angle θ about the z-axis, the matrix $\mathbf{R}$ has the familar form

$$\mathbf{R} = \begin{pmatrix} \cos\theta & \sin\theta & 0 \\ -\sin\theta & \cos\theta & 0 \\ 0 & 0 & 1 \end{pmatrix}. \tag{A.3}$$

Pictorially, we have Fig. A.2.

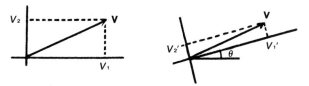

Fig. A.2. Transformation of vector under rotation.

It is customary to use the so-called *summation convention* in writing out such quantities. The convention can be stated as follows: Whenever an index is repeated, it is understood that there is a summation over that index, with the index itself running from 1 to 3. Using the summation convention, Eq. (A.2) can be written

$$V'_i = R_{ij} V_j. \tag{A.4}$$

Up to this point, we have proceeded as if we knew what a vector was, and were deriving the law which told us how that vector appeared in different frames of reference. We can, however, reverse the logic, and use Eq. (A.4) as a *definition* of a vector—i.e. we define a vector as any collection of three numbers which transforms according to the law in Eq. (A.4). This concept of defining an object by the way in which it changes under coordinate transformations is a fairly recent development in physics, and has been enormously useful in fields as widely separated as nuclear physics and the general theory of relativity.

If we take this point of view, we see that it is possible to construct other kinds of objects. For example, suppose we define as a second-rank tensor any object with two indices (i.e. 9 components) which transforms according to the law

$$T'_{ij} = R_{il} R_{jm} T_{lm}. \tag{A.5}$$

An example of such an object would be the tensor whose $i - j$th component is

$$T_{ij} = V_i V_j,$$

where V is a vector.

To see that this quantity satisfies the definition of a tensor, we note that in one frame, $T_{ij} = V_i V_j$, while in the frame which has been rotated

$$T'_{lm} = V'_l V'_m, \tag{A.6}$$

but

$$V'_l = R_{li} V_i, \tag{A.7}$$

$$V'_m = R_{mj} V_j,$$

so that the tensor in the rotated frame is

$$T'_{lm} = R_{li} R_{mj} V_i V_j = R_{li} R_{mj} T_{ij}, \tag{A.8}$$

where T_{ij} is the tensor in the old frame. Other examples of such tensors will be found in the text.

It must be noted that not every two index object is a tensor, and if something is to be called a tensor, it must be explicitly verified that it transforms according to the transformation law of Eq. (A.5).

It should also be noted that in the nomenclature introduced above, a vector could be referred to as a first-rank tensor. Clearly, third-, fourth-, and higher-rank tensors can be defined in complete analogy to the definition in Eq. (A.5).

The greatest use which we shall make of the Cartesian tensor notation will not be concerned with second-rank tensors, however, but shall be the utilization of the very compact and efficient notation it provides for manipulating vectors and vector operators. Often operations which appear quite complicated when written in vector form are simple to analyze in terms of tensor notation.

Let us therefore catalogue several common vector operations in both vector and tensor form.

(A) Inner Product

The inner product between two vectors is just

$$\mathbf{A} \cdot \mathbf{B} = A_x B_x + A_y B_y + A_z B_z = A_i B_i. \tag{A.9}$$

This can also be written as

$$\mathbf{A} \cdot \mathbf{B} = A_i B_j \delta_{ij}, \tag{A.10}$$

where δ_{ij}, the Kronecker delta, is defined by

$$\delta_{ij} = \begin{cases} 1 & i = j \\ 0 & i \neq j \end{cases}. \tag{A.11}$$

(B) Gradient

The gradient of a function f is defined to be

$$\mathbf{D}f = \frac{\partial f}{\partial x}\,\hat{i} + \frac{\partial f}{\partial y}\,\hat{j} + \frac{\partial f}{\partial z}\,\hat{k}, \tag{A.12}$$

where $\hat{i}$, $\hat{j}$, and $\hat{k}$ are unit vectors in the x-, y-, and z-directions. In tensor notation, this becomes

$$(\nabla f)_i = \frac{\partial f}{\partial x_i}. \tag{A.13}$$

(C) Divergence

The divergence of a vector is

$$\nabla \cdot \mathbf{A} = \frac{\partial A_x}{\partial x} + \frac{\partial A_y}{\partial y} + \frac{\partial A_z}{\partial z}$$

which can be written

$$\nabla \cdot \mathbf{A} = \frac{\partial A_i}{\partial x_i}. \tag{A.14}$$

(D) Cross Product

The cross product of a vector can be written

$$\mathbf{A} \times \mathbf{B})_i = \epsilon_{ijk} A_j B_k, \tag{A.15}$$

where ϵ_{ijk} is defined by

$$\epsilon_{ijk} = \begin{cases} +1 & i, j, k \text{ cyclic} \\ -1 & i, j, k \text{ anti-cyclic} \\ 0 & \text{any 2 indices equal} \end{cases}. \tag{A.16}$$

(E) Curl

The curl is a special case of the cross product and is written

$$\nabla \times \mathbf{A})_i = \epsilon_{ijk} \frac{\partial}{\partial x_j}\,A_k. \tag{A.17}$$

APPENDIX B THE GRAVITATIONAL POTENTIAL INSIDE OF A UNIFORM ELLIPSOID

In this appendix, we will work through the straightforward but tedious derivation of the gravitational potential inside of a uniform ellipsoid. This quantity is necessary for the study of classical stellar structure in Chapter 2.

Consider a point P, whose coordinates are x_P, y_P, and z_P inside an ellipsoid whose surface is described by the equation

$$\frac{x^2}{a^2} + \frac{y^2}{b^2} + \frac{z^2}{c^2} = 1. \tag{B.1}$$

We begin by changing coordinates (see Fig. B.1) to

$$
\begin{aligned}
x &= x_P + r \sin\theta \cos\phi, \\
y &= y_P + r \sin\theta \sin\phi, \\
z &= z_P + r \cos\theta.
\end{aligned}
\tag{B.2}
$$

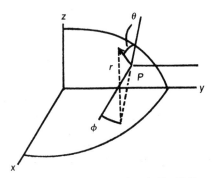

Fig. B.1.   Coordinates given in Eq. (B.2).

The volume element in the new variables is the usual one for spherical coordinates

$$dV = r^2 \, dr \, d(\cos\theta) \, d\phi, \tag{B.3}$$

so that the gravitational potential at P is now just

$$\Omega_p = G\rho \int \frac{dV}{r} = G\rho \int_{-\pi/2}^{\pi/2} \int_0^{2\pi} \int_0^{r_1} r \, dr \, d(\cos\theta) \, d\phi, \tag{B.4}$$

where r_1 is the distance from P to the boundary of the ellipsoid for a given choice of θ and ϕ (clearly, r_1 will be a function of both angular variables and of P).

To solve for r_1, it is necessary only to put the values of expressions for x, y, and z from Eq. (B.2) into the equation describing the boundary, Eq. (B.1). We quickly find

$$A r_1^2 + 2B r_1 + C = 0, \tag{B.5}$$

where

$$A = \frac{\sin^2 \theta \cos^2 \phi}{a^2} + \frac{\sin^2 \theta \sin^2 \phi}{b^2} + \frac{\cos^2 \theta}{c^2},$$

$$B = \frac{x_P \sin \theta \cos \phi}{a^2} + \frac{y_P \sin \theta \sin \phi}{b^2} + \frac{z_P \cos \theta}{c^2}, \tag{B.6}$$

$$C = \frac{x_P^2}{a^2} + \frac{y_P^2}{b^2} + \frac{z_P^2}{c^2} - 1.$$

It is a simple exercise to show that for points inside the body, the correct choice of signs in the quadratic formula gives

$$r_1 = \frac{-B + \sqrt{B^2 - AC}}{A}, \tag{B.7}$$

so that after performing the integral over the r-coordinate

$$\Omega = \frac{G\rho}{2} \int_{-\pi/2}^{\pi/2} \int_0^{2\pi} d(\cos \theta) \, d\phi \, \frac{2B^2 - AC - 2B\sqrt{B^2 - AC}}{A^2}. \tag{B.8}$$

In principle, we could now simply substitute the definitions of A, B, and C from Eq. (B.6) into this integral and carry out the integrations. However, we can note several symmetries in the integrand which greatly simplify the result.

First, we note that if we let

$$\phi \to \pi + \phi,$$
$$\phi \to \pi - \theta.$$

A and C remain unchanged, but B goes to $-B$. Thus, in integrating over the complete solid angle, terms linear in B will give a zero integral. Thus, the term involving the radical in the integrand above can be dropped.

Similarly, in calculating the term involving B^2, we expect that there will be terms proportional to x_P^2, y_P^2, and z_P^2, and, in addition, cross terms proportional to x_P, y_P, etc. Arguments similar to that in the preceding paragraph can be evoked to show that the cross terms do not contribute the final result. Consider as an example the term

$$\frac{x_P y_P}{a^2 b^2} \sin^2 \theta \cos \phi \sin \phi.$$

This term will change sign under the transformation $\phi \to -\phi$, and hence will vanish when integrated over the solid angle. Similar arguments can be made for the other cross terms.

We are then left with

$$\Omega_P = G_P \int_{-\pi/2}^{\pi/2} \int_0^{2\pi} d(\cos\theta)\, d\phi \left[x_P^2 \frac{\sin^2\theta \cos^2\phi}{a^2} + y_P^2 \frac{\sin^2\theta \sin^2\phi}{b^4} \right.$$
$$\left. + z_P^2 \frac{\cos^2\theta}{c^4} \right] \frac{1}{A^2} - \frac{1}{2} G\rho C \int_{-\pi/2}^{\pi/2} \int_0^{2\pi} \frac{d(\cos\theta)\, d\phi}{A}. \tag{B.9}$$

These integrals are still rather complicated, but there is a trick which will allow us to put them into much simpler form. Let us define the quantity

$$W = \frac{G\rho}{2} \int \int \frac{d(\cos\theta)\, d\phi}{A}. \tag{B.10}$$

Then it is simple to show that

$$\frac{\partial W}{\partial a} = \frac{G\rho}{2} \int \frac{d(\cos\theta)\, d\phi}{A^2} \left[\frac{\sin^2\theta \cos^2\phi}{a^3} \right]. \tag{B.11}$$

Similar expressions can be written for $\partial W/\partial b$ and $\partial W/\partial c$. If we put all of these into the above integral, and recall the definition of C, we find

$$\Omega_P = \left(\frac{1}{a} \frac{\partial W}{\partial a} - \frac{W}{a^2} \right) x_P^2 + \left(\frac{1}{b} \frac{\partial W}{\partial b} - \frac{W}{b^2} \right) y_P^2$$
$$+ \left(\frac{1}{c} \frac{\partial W}{\partial c} - \frac{W}{c^2} \right) z_P^2 + W \tag{B.12}$$
$$= \alpha x_P^2 + \beta y_P^2 + \gamma z_P^2 + \chi,$$

which is the general form which we used in Chapter 2. To get our final result, we have only to evaluate W.

Actually, this cannot be done in closed form for an arbitrary ellipsoid, but we can carry out one of the angular integrals in the definition of W by making the substitution

$$M = \frac{\sin^2\theta}{a^2} + \frac{\cos^2\theta}{c^2},$$
$$N = \frac{\sin^2\theta}{b^2} + \frac{\cos^2\theta}{c^2}. \tag{B.13}$$

It is then possible to carry out the integral over ϕ by writing

$$W = \frac{G\rho}{2} \int_{-\pi/2}^{\pi/2} d(\cos\theta) \int_0^{2\pi} \frac{d\phi}{M\cos^2\phi + N\sin^2\phi}$$
$$= 2\pi\rho G \int_{-\pi/2}^{\pi/2} \frac{d(\cos\theta)}{\sqrt{MN}}, \tag{B.14}$$

which can be put into a somewhat more familiar form by changing variables to λ, where

$$\sin \theta = \frac{c}{\sqrt{c^2 + \lambda}},$$

to give

$$W = \pi\rho Gabc \int_0^\infty \frac{d\lambda}{\sqrt{(a^2 + \lambda)(b^2 + \lambda)(c^2 + \lambda)}}. \tag{B.15}$$

APPENDIX C THE CRITICAL FREQUENCY

In Chapter 2, we saw that for a Maclaurin ellipsoid, it was impossible to achieve equilibrium if the frequency of rotation exceeded a certain value, on the order of the critical frequency

$$\omega_c^2 = 2\pi\rho G. \tag{C.1}$$

In this appendix, we will show that the critical frequency is the upper limit on the frequency of rotation for any incompressible body.

Consider an arbitrary volume V in the rotating fluid, surrounded by a surface S. If ϕ and ψ are any two functions, then *Green's theorem* tells us that

$$\int_V [\phi\nabla^2\psi + \psi\nabla^2\phi]\, dV = \int_S \left[\phi\frac{\partial\psi}{\partial n} - \psi\frac{\partial\phi}{\partial n}\right] dS, \tag{C.2}$$

where $\partial/\partial n$ represents the derivative of the function along the outward normal to S.

Now let us take the case

$$\phi = 1$$

and

$$\psi = P.$$

Then Eq. (C.2) becomes

$$\int_V \nabla^2 P\, dV = \int_S \frac{\partial P}{\partial n}\, dS. \tag{C.3}$$

If we substitute in the left-hand integral the expression for P which we obtained by integrating the Euler equation (2.A.4), we have

$$\int_V \nabla^2 P\, dV = \int_V \nabla^2 \left\{ \left[\frac{\rho\omega^2}{2}(x^2 + y^2)\right] - \rho\Omega \right\} dV$$
$$= 2\rho V(\omega^2 - 2\pi\rho G), \tag{C.4}$$

where V is the total volume and we have used the expression

$$\nabla^2\Omega = 4\pi\rho G \tag{C.5}$$

to eliminate the gravitational potential.

The pressure on the surface of our body is constant, and for purposes of explanation, we can take it to be zero. If $\omega > \omega_c$, then the integral of $\int \partial P / \partial n \, ds$ over any surface in the fluid must be positive, which means that the pressure must be increasing as we go from the interior of the fluid toward the surface. Thus, the pressure forces act inward, in the same direction as gravity. Equilibrium in such a case is clearly impossible, since the forces in the z-direction on any element of fluid will not cancel each other (we talk about the z-direction because the centrifugal force has no z-component).

On the other hand, if $\omega < \omega_c$, the pressure must decrease as we move from the center of the fluid to the surface, and it is possible for equilibrium to be achieved. Whether or not this possibility is actually realized depends, of course, on the shape of the fluid mass.

Thus, we see that on very general grounds, no fluid mass can be in equilibrium if it is spinning with a frequency greater than ω_c, which is what we set out to prove.

APPENDIX D EXPANSION IN ORTHOGONAL POLYNOMIALS

Throughout the text, we have used the idea of expanding arbitrary functions in terms of other, simpler functions. In this appendix, we will discuss this idea in detail, although for a rigorous proof of the things we say, the reader will have to consult a mathematics textbook.

The idea of expansion is actually a familiar one. Consider a vector $\mathbf{V}$ in a Cartesian coordinate system as shown in Fig. D.1. We know that we can always expand this vector in terms of the three basis vectors, $\hat{i}$, $\hat{j}$, and $\hat{k}$

$$\mathbf{V} = V_x \hat{i} + V_y \hat{j} + V_z \hat{k}, \tag{D.1}$$

where the components V_i are given by

$$V_x = \mathbf{V} \cdot \hat{i}, \qquad V_y = \mathbf{V} \cdot \hat{j}, \qquad V_z = \mathbf{V} \cdot \hat{k}.$$

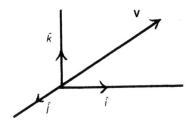

Fig. D.1. A vector in three dimensions.

The basis vectors have two important properties. First, they are orthogonal to each other, so that

$$\hat{i} \cdot \hat{j} = \hat{i} \cdot \hat{k} = \hat{j} \cdot \hat{k} = 0, \tag{D.2}$$

and second, they are normalized, so that

$$\hat{i} \cdot \hat{i} = \hat{j} \cdot \hat{j} = \hat{k} \cdot \hat{k} = 1. \tag{D.3}$$

A set of vectors which has these properties is called an *orthonormal set* of vectors.

We can use a slightly different notation in writing down these facts about expanding a vector in terms of its components. If we denote by $\hat{\theta}_i$ the basis vector in the ith-direction, then the requirement of orthonormality takes the form

$$\hat{\theta}_i \cdot \hat{\theta}_j = \delta_{ij}, \tag{D.4}$$

while the expansion of the vector $\mathbf{V}$ can be written

$$\mathbf{V} = \sum_{i=1}^{3} (\mathbf{V} \cdot \hat{\theta}_i)\hat{\theta}_i = \sum_{i=1}^{3} a_i \hat{\theta}_i. \tag{D.5}$$

In what follows, we will call the constant a_i the coefficient of expansion.

Now there is nothing in the above development which forces us to confine our attention to three-dimensional spaces. If we considered a vector $\mathbf{V}$ in an N-dimensional space, and defined a set of basis vectors $\hat{\theta}_i$ as in Eq. (D.4), but now let the index i run up to N rather than just to 3, then we could expand the new vector as

$$\mathbf{V} = \sum_{i=1}^{N} (\mathbf{V} \cdot \hat{\theta}_i)\hat{\theta}_i \tag{D.6}$$

by simple analogy.

Consider now a function $f(x)$ defined on some interval in x, say from zero to L (see Fig. D.2). Let us split the interval up into N equal spaces

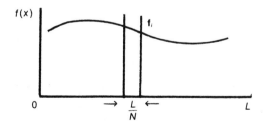

Fig. D.2.   The representation of a function by a vector.

and form an N-dimensional vector

$$\mathbf{F} = (\mathbf{f}_1, \mathbf{f}_2, \ldots, \mathbf{f}_i), \tag{D.7}$$

where $\mathbf{f}_N$ is the average value of the function $f(x)$ in the ith interval multiplied by $\sqrt{L/N}$. In exactly the same way, we could form a vector

$$\mathbf{G} = (\mathbf{g}_1 \ldots \mathbf{g}_N), \tag{D.8}$$

from the function $g(x)$ defined on the same interval. The inner product in the N-dimensional space between the vectors $\mathbf{G}$ and $\mathbf{F}$ is just

$$\mathbf{G} \cdot \mathbf{F} = \sum_{j=1}^{N} \mathbf{f}_j \cdot \mathbf{g}_j. \tag{D.9}$$

Suppose now that we were able to find a set of functions $\phi^{(\alpha)}(x)$ defined on x from zero to L, as were $f(x)$ and $g(x)$, and were to form a vector $\phi^{(\alpha)}$ as in Eq. (D.7) for each of these new functions. Suppose also that the vectors so formed had the property that

$$\phi^{(\alpha)} \cdot \phi^{(\beta)} = \sum_{i=1}^{N} \phi_i^{(\alpha)} \phi_i^{(\beta)} = \delta_{\alpha\beta}. \tag{D.10}$$

(It is important to distinguish between the superscript α in $\phi_i^{(\alpha)}$ and the subscript j. The superscript refers to the index which tells us which function we are discussing, while the subscript tells us which interval in x is being considered.) Then the vectors formed in this way would be an orthonormal set of basis vectors and we could write

$$\mathbf{F} = \sum_{\alpha} (\mathbf{F} \cdot \phi^{(\alpha)}) \phi^{(\alpha)} = \sum_{\alpha} \sum_{j=1}^{N} (F_j \phi_j^{(\alpha)}) \phi^{(\alpha)}, \tag{D.11}$$

and similarly for $\mathbf{G}$.

It should be emphasized that up to this point, no new information has been presented, and we have only been presenting consequences of the known properties of vectors. Let us ask what happens, however, if we let N go to infinity. In this case, the number of components in the vector defined in Eq. (D.7) becomes infinite and the sum over indices in the inner product in Eq. (D.9) gets converted to an integral, so that our new definition of the inner product between the new vectors becomes

$$\mathbf{F} \cdot \mathbf{G} = \int_0^L f(x)g(x)\, dx. \tag{D.12}$$

What we have done is define a new vector space of infinite dimension, in which each function $f(x)$ is represented by a vector. This is an example of a *Hilbert space*. For the sake of honesty, it must be pointed out that this is simply an analogy that we have drawn here, and the reader wishing more rigor in the definition of these spaces is referred to texts in mathematics.

Suppose in our Hilbert space the basis functions retain their orthonormality, so that

$$\int_0^L \phi^{(\alpha)}(x)\phi^{(\beta)}(x)\,dx = \delta_{\alpha\beta}. \tag{D.13}$$

Then the analogue of Eq. (D.11) is just

$$\begin{aligned}
f(x) &= \sum_\alpha (f \cdot \phi^{(\alpha)})\phi^{(\alpha)}(x) \\
&= \sum_\alpha \left[\int_0^L f(x)\phi^{(\alpha)}(x)\,dx \right] \phi^{(\alpha)}(x).
\end{aligned} \tag{D.14}$$

Thus, by analogy to the expansion of an ordinary vector in terms of its basis vectors, we can expand an arbitrary function in terms of a set of basis functions which satisfy Eq. (D.13).

Do such sets of basis vectors exist? The answer to this question is yes—there are, in fact, many such sets. Consider, for example, the set of functions

$$\phi^{(n)}(x) = \sqrt{\frac{2}{L}} \sin \frac{n\pi x}{L},$$

$$\psi^{(n)}(x) = \sqrt{\frac{2}{L}} \cos \frac{n\pi x}{L}, \tag{D.15}$$

defined on the interval $0 \le x \le L$. Some simple calculations will convince the reader that

$$\phi^{(n)} \cdot \phi^{(m)} = \psi^{(n)} \cdot \psi^{(m)} = \delta_{m,n}$$

and

$$\phi^{(n)} \cdot \psi^{(m)} = 0. \tag{D.16}$$

Thus, the sines and cosines form a set of basis vectors in a Hilbert space, just as the vectors $\hat{i}$, $\hat{j}$, and $\hat{k}$ form a *complete set*—i.e. that there is no vector in the Hilbert space orthogonal to all the $\phi^{(n)}$ and $\psi^{(n)}$, just as there is no vector in Cartesian space orthogonal to $\hat{i}$, $\hat{j}$, and $\hat{k}$.

This means that any function defined on the interval $0 \le x \le L$ can be written in the form

$$f(x) = \sum a_n \sin\left(\frac{n\pi x}{L}\right) + \sum b_n \cos\left(\frac{n\pi x}{L}\right), \tag{D.17}$$

where

$$a_n = \sqrt{\frac{2}{L}} \int_0^L f(x) \sin\frac{n\pi x}{L}\, dx$$

and

$$b_n = \sqrt{\frac{2}{L}} \int_0^L f(x) \cos\left(\frac{n\pi x}{L}\right) dx.$$

An expansion of this type is called a *Fourier series*, and plays an extremely important role in physics. The reader will see, however, that it is simply one example of an expansion of a function in orthogonal polynomials, and if we can find another set of functions like those in Eq. (D.15), alternate series representations of the function will be possible, just as a three-dimensional vector can be expanded in Cartesian, spherical, or cylindrical coordinates. Further examples of orthonormal basis sets are given in Appendix F.

APPENDIX E   SOLUTION OF ORDINARY DIFFERENTIAL EQUATIONS

There is no "right way" or general method to solving differential equations. It is an art, rather than a science. By this I mean that the solution of differential equations involves making educated guesses at solutions, rather than proceeding by logical steps from some set of first principles. In this appendix, we will review the most common forms of solutions to ordinary linear equations, and discuss some important properties of the solutions.

The most general equation of this type is

$$f_n(x)y^{(n)}(x) + \cdots f_0(x)y(x) = g(x), \tag{E.1}$$

where $y(x)$ is a function which is to be determined, $y^{(n)}(x)$ is the nth derivative, $f_0(x)\ldots f_n(x)$ are known functions of x, and $g(x)$, the inhomogeneous term, is also known.

In the text, we most often considered equations of second order, i.e. equations where $n = 2$. Let us begin by considering the homogeneous equation of order 2, which is

$$f_2(x)\frac{d^2y}{dx^2} + f_1(x)\frac{dy}{dx} + f_0(x)y = 0. \tag{E.2}$$

The general method of solving such an equation is to guess a form of solution, and then see if that form can be made to fit the equation. For example, we might guess a solution for $y(x)$ in Eq. (E.2) to be of the form

$$y(x) = Ce^{\alpha x}. \tag{E.3}$$

Let us consider only equations where

$$f_n(x) = 1$$

and

$$f_1(x) = C_1, \qquad f_2(x) = C_2.$$

Then substituting Eq. (E.3) into Eq. (E.2) gives an equation

$$Ae^{\alpha x}[\alpha^2 + C_1\alpha + C_2] = 0, \tag{E.4}$$

which can be solved for α. In general, there will be solutions of the form

$$\alpha = \beta \pm \gamma \tag{E.5}$$

from the quadratic formula, where

$$\beta = -\tfrac{1}{2}C_1,$$

$$\gamma = \tfrac{1}{2}\sqrt{C_1^2 - 4C_2}.$$

The constant A cannot be determined from the equation, of course.

We are now in a position in which we have two possible solutions of the form (E.3). One is

$$y_1(x) = A_1 e^{(\beta + \gamma)y},$$

while the other is

$$y_2(x) = A_2 e^{(\beta - \gamma)y},$$

where A_1 and A_2 are arbitrary constants.

What is the most general solution to Eq. (E.2)? If we substitute the form

$$y(x) = y_1(x) + y_2(x) \tag{E.6}$$

into Eq. (2), we see that it, too, is a solution of the equation. It is, in fact, the most general solution to the equation (the proof of this is left to textbooks in mathematics). The general theorem (which can easily be proved by simple substitution) is that if

$$\phi_1, \phi_2, \ldots, \phi_n$$

are solutions to a general nth-order homogeneous equation, then the most general solution will be

$$\phi = A_1\phi_1 + A_2\phi_2 + \cdots A_n\phi_n, \tag{E.7}$$

where A_n are arbitrary constants.

The constants A_1 and A_2 cannot, as we have seen, be determined from the equation alone, but must be derived from additional information. This information is usually given in the form of boundary conditions. There are many examples of this in the text. For example, we might be given the value of $y(x)$ at two points, or the value of $y(x)$ and dy/dx at a single point. As long as we have two boundary conditions (or n conditions for the nth-order equation), we can determine the arbitrary constants, and thereby fix the solution exactly. It must be emphasized that boundary conditions are generally given by consideration of the physics of the situation, rather than the mathematics.

Let us now turn our attention to the more general form of Eq. (E.2), namely the inhomogeneous equation of order 2

$$f_2(x)\frac{d^2y}{dx^2} + f_1(x)\frac{dy}{dx} + f_0(x)y = g(x). \tag{E.8}$$

There are several things which we can say about this equation. First of all, suppose that $y_p(x)$ is a solution of Eq. (E.8). Then simple substitution shows that

$$y(x) = y_p(x) + y_h(x) \tag{E.9}$$

is also a solution of Eq. (E.8) provided that $y_h(x)$ is a solution of Eq. (E.2). Thus, we see that to any particular solution of Eq. (E.8), which we have called y_p, we can add any solution or combinations of solutions of the homogeneous equation. Thus, there are just as many undetermined constants in the inhomogeneous equation as there were in the homogeneous, and they, too, must be determined from the boundary conditions.

How can the particular solution y_p be found? Once again there are no general procedures, but we have to make a guess, and then see if it will work. For example, take the equation

$$\frac{d^2y}{dx^2} + C_1\frac{dy}{dx} + C_2y = C_3. \tag{E.10}$$

Then a reasonable guess might be

$$y_p(x) = F,$$

where F is a constant. Substituting this guess back into Eq. (E.10), we find that it will satisfy the equation provided that

$$F = \frac{C_3}{C_2}.$$

Thus, the most general solution to Eq. (E.10) is just

$$y_G(x) = F + y(x), \tag{E.11}$$

where $y(x)$ is given in Eq. (E.6).

There is one important property of the inhomogeneous equation which we have used throughout the text. Consider the inhomogeneous equation of the form

$$f_2(x)\frac{d^2y}{dx^2} + f_1(x)\frac{dy}{dx} + f_0(x)y = A_1g_1(x) + A_2g_2(x), \tag{E.12}$$

and let y_{1p} be a solution of

$$f_2(x)\frac{d^2y}{dx^2} + f_1(x)\frac{dy}{dx} + f_0(x)y = A_1g_1(x), \tag{E.13}$$

while y_{2p} is a solution of

$$f_2(x)\frac{d^2y}{dx^2} + f_1(x)\frac{dy}{dx} + f_0(x)y = A_2g_2(x). \tag{E.14}$$

Thus, by substitution, we can see that the most general solution of Eq. (E.12) will just be

$$y(x) = y(x) + y_{1p}(x) + y_{2p}(x). \tag{E.15}$$

The generalization on this statement to any number of terms on the right-hand side is obvious. Given the method of expansion in orthogonal polynomials discussed in Appendix D, we can always write the inhomogeneous term in Eq. (E.8) as

$$g(x) = \sum a_n\theta_n(x),$$

where $\theta_n(x)$ is some suitable set of orthogonal polynomials. For example, if we were expanding in a Fourier series, we would have

$$g(x) = \sum a_n \sin\frac{n\pi x}{L} + \sum b_n \cos\frac{n\pi x}{L}.$$

By the theorem stated above, however, if we wished to solve this equation, it would be sufficient to solve the equation

$$f_2(x)\frac{d^2y}{dx^2} + f_1(x)\frac{dy}{dx} + f_0(x)y = a_n \sin\frac{n\pi x}{L}. \tag{E.16}$$

The general solution to Eq. (E.8) would then be

$$y(x) = y_n(x) + \sum_n y_{pn}(x), \qquad (E.17)$$

where $y_{pn}(x)$ is the solution to Eq. (E.16). This is the mathematical basis behind the general procedure we follow in the text of solving for one Fourier component only, and neglecting all others.

APPENDIX F   THE SOLUTION OF PARTIAL DIFFERENTIAL EQUATIONS

Although partial differential equations, like ordinary equations, cannot necessarily be solved by applying some general method, a large class of the equations which are of most interest to physicists can be solved by the technique known as the *separation of variables*. This technique is, like the methods discussed in Appendix D, a guess at what the solution of an equation will look like. The technique is to guess a solution, plug it into the equation, and see if it works. If it does, then we know from the theory of differential equations that the solution is unique.

We shall take as our example the equation which describes potential flow of an incompressible fluid,

$$\nabla^2 \phi = 0, \qquad (F.1)$$

which is called Laplace's equation. In Cartesian coordinates this becomes

$$\frac{\partial^2 \phi}{\partial x^2} + \frac{\partial^2 \phi}{\partial y^2} + \frac{\partial^2 \phi}{\partial z^2} = 0. \qquad (F.2)$$

The essential step in the technique is to assume that the solution is of the form

$$\phi = X(x)\, Y(y)\, Z(z), \qquad (F.3)$$

where $X(x)$ is a function of x only, $Y(y)$ of y, etc. If we put this assumed form into Eq. (F.2) and divide by XYZ, we find

$$\frac{X''}{X} + \frac{Y''}{Y} = \frac{Z''}{Z} = 0,$$

where the prime denotes differentiation with respect to the argument of the function.

Now what we have here is a situation in which a function of x alone must be equal to the negative of a function of y and z alone. The only way that this can be true for all values of x is for that function to be a constant, so that

$$\frac{X''}{X} = \alpha^2. \tag{F.4}$$

Similarly,

$$\frac{Y''}{Y} = \beta^2$$

and

$$\frac{Z''}{Z} = -\alpha^2 - \beta^2.$$

We have now reduced the problem of solving a partial differential equation to the problem of solving three ordinary (and in this case identical) equations. From Appendix E, we know that the solution to these equations is of the form

$$X = A \sin \alpha x + B \cos \alpha x, \tag{F.5}$$

with similar forms for Y and Z. The actual determination of the constants A and B, as in Appendix D, is done by applying the boundary conditions. The determination of the constants α and β is also done by applying the boundary conditions, but in a somewhat more subtle way.

Suppose that the fluid we are considering is confined to a cube of side L. Suppose further that we know that the potential along some line of constant y and z is given by a known function $F(x)$. Then we must have

$$\phi(x, y_1, z_1) = F(x) = X(x)Y(y_1)Z(z_1), \tag{F.6}$$

where y_1 and z_1 are the values of y and z along the plane. Since $Y(y_1)Z(z_1)$ is just a constant, we must have that the solution for $X(x)$ in Eq. (F.5) reduces to $F(x)$ in this case.

From Appendix D and the definition of a Fourier series, we know that we can do this by choosing the constants such that

$$\alpha = \frac{n\pi}{L}$$

and

$$A = a_n = \frac{2}{L} \int_0^L F(x) \sin \frac{n\pi x}{L} \, dx,$$

$$B = b_n = \frac{2}{L} \int_0^L F(x) \cos \frac{n\pi x}{L} \, dx,$$

so that, up to an overall constant,

$$X(x) = \sum a_n \sin\frac{n\pi x}{L} + \sum b_n \cos\frac{n\pi x}{L}. \tag{F.7}$$

This is the solution of a typical *boundary-value problem*. In general, $Y(y)$ and $Z(z)$ will also be given by Fourier series of some boundary-value functions. The reader is referred to the texts at the end of the introduction to the appendices for more detailed discussion of this point. The main thing that we want to emphasize is that the solution of the Laplace equation is intimately tied to the existence of orthonormal sets of polynomials (in this case the sines and cosines) which are, in fact, the solutions to the ordinary differential equations which result from applying the technique of separation of variables.

We might guess, then, that other sets of these polynomials might arise from solutions of the equation in other coordinate systems. For example, just as the sines and cosines are particularly appropriate for expanding functions in Cartesian coordinates, there might be other functions which are appropriate for expanding functions in spherical coordinates.

The Laplace equation in spherical coordinates is

$$\frac{1}{r^2}\frac{\partial}{\partial r}\left(r^2\frac{\partial\Phi}{\partial r}\right) + \frac{1}{r^2\sin\theta}\frac{\partial}{\partial\theta}\left(\sin\theta\frac{\partial\Phi}{\partial\theta}\right) + \frac{1}{r^2\sin^2\theta}\frac{\partial^2\Phi}{\partial\phi^2} = 0. \tag{F.8}$$

If we proceed as before and assume a solution of the form

$$\Phi = R(r)P(\theta)Q(\phi), \tag{F.9}$$

then we find, upon dividing by $\Phi r^2\sin^2\theta$,

$$\frac{1}{R}\sin^2\theta\frac{\partial}{\partial r}\left(r^2\frac{\partial R}{\partial r}\right) + \frac{\sin\theta}{P}\frac{\partial}{\partial\theta}\left(\sin\theta\frac{\partial P}{\partial\theta}\right) + \frac{1}{Q}\frac{\partial^2 Q}{\partial\phi^2} = 0.$$

As before, we note that this can only be true if the function of ϕ is a constant, which we take to be

$$\frac{1}{Q}\frac{\partial^2 Q}{\partial\phi^2} = -m^2,$$

so that

$$Q = e^{\pm im\phi}. \tag{F.10}$$

Clearly, if we wish the function to be single valued, so that

$$\Phi(r,\,\theta,\,\phi) = \Phi(r,\,\theta,\,\phi + 2\pi),$$

we must have m be an integer (this is actually the first application of the boundary conditions). We are then left with

$$\frac{1}{R}\frac{\partial}{\partial r}\left(r^2\frac{\partial R}{\partial r}\right) + \frac{1}{P\sin\theta}\frac{\partial}{\partial\theta}\left(\sin\theta\frac{\partial P}{\partial\theta}\right) - \frac{m^2}{\sin^2\theta} = 0.$$

Once more, this equation can be satisfied only if the function of R is a constant, which, for convenience, we will take to be

$$\frac{1}{R}\frac{\partial}{\partial r}\left(r^2\frac{\partial R}{\partial r}\right) = l(l+1),$$

where l can be any number. From the methods of Appendix E, assuming an R of the form r^q yields as a solution

$$R(r) = Ar^l + \frac{B}{r^{l+1}}, \tag{F.11}$$

and leaves us with the result

$$\frac{1}{\sin\theta}\frac{d}{d\theta}\left(\sin\theta\frac{dP}{d\theta}\right) + \left[l(l+1) - \frac{m^2}{\sin^2\theta}\right] = 0. \tag{F.12}$$

This is known as *Legendre's equation.* It is usually written in a form where the change of variables

$$x = \cos\theta$$

has been made, so that

$$\frac{d}{dx}\left[(1-x^2)\frac{dP}{dx}\right] + \left[l(l+1) - \frac{m^2}{1-x^2}\right]P = 0. \tag{F.13}$$

Let us consider first the case of azimuthal symmetry, where $m = 0$. The equation to be solved is then

$$\frac{d}{dx}\left[(1-x^2)\frac{dP}{dx}\right] + l(l+1)P = 0. \tag{F.14}$$

Let us assume that we can find a solution of this equation of the form

$$P = \sum_{n=0}^{\infty} a_n x^n, \tag{F.15}$$

where the coefficients a_n are to be determined. Provided that everything is well behaved, this is not a large assumption, since it amounts to expanding the solution in a Taylor series.

If we insert this assumed form of solution into Eq. (F.14), we find

$$[l(l+1)]c_0 + 2c_o + [(l(l-1)-2)c_1 + 6c_3]x$$
$$+ [(l(l+1)-6)c_2 + 12c_4]x^2 + \cdots = 0.$$

Now in order for there to be a solution which is valid for every value of x, the coefficient of each power of x must vanish identically. This means that

$$c_2 = -\frac{l(l+1)}{2}c_0$$

and

$$c_4 = \frac{6 - l(l+1)}{12}c_2,$$

with similar relations between c_1, c_3, c_5, etc. In general, we have

$$\frac{c_{i+2}}{c_i} = \frac{(i+l-i)(i-l)}{(i+1)(i+2)}. \tag{F.16}$$

There are several points to note about this result. First of all, if any c_n is ever zero, then every higher value of n will also have a vanishing coefficient. For example, if c_6 were zero, then Eq. (F.16) would give c_8 to be zero, and applying the equation again would give $c_{10} = 0$, and so forth. A second point is that every term with even n can be related back to c_0 by repeated use of Eq. (F.16), and every term of odd n can be related back to c_1. Furthermore, the odd and even terms are not related to each other so that Eq. (F.15) can be written

$$P(x) = c_0 \sum a_{2n}x^{2n} + c_1 \sum b_{2n+1}x^{2n+1}, \tag{F.17}$$

i.e. as a sum of even indices plus a sum over odd indices. There is nothing in the equation, however, to tell us what to take for c_1 and c_0. By convention, we usually take either c_0 or c_1 to be zero (so that the solution is either odd or even), and adjust the nonzero coefficient such that

$$P(0) = 1.$$

The polynomials which are generated in this way are called the *Legendre polynomials*. We note that if l is an integer, then the factor $i - l$ in Eq. (F.16) will vanish when $i = l$, so that the polynomial will be of order l, and will contain no higher powers of x. For this reason, it is customary to denote the Legendre polynomial by P_l. The first few polynomials are

$$P_0 = 1,$$
$$P_1 = x,$$
$$P_2 = \frac{1}{2}(3x^2 - 1),$$
$$P_3 = \frac{1}{2}(5x^3 - 3x),$$

and higher orders can be worked out from the recursion relation in Eq. (F.16).

Having solved Laplace's equation in spherical coordinates, we now ask ourselves whether the solutions in this case form an orthonormal set, as did the sines and cosines in the Cartesian case. It is relatively simple to show that

$$\int P_l(x)P_{l'}(x)\,dx = \frac{2}{2l+1}\delta_{ll'} \tag{F.18}$$

[see, for example, the text by Mathews and Walker cited in the bibliography]. Therefore, if we define

$$U_l(x) = \sqrt{\frac{2l+1}{2}}P_l(x), \tag{F.19}$$

then the $U_l(x)$ form an orthonormal set for expansion of functions as series in cos θ, just as the sines and cosines did for expansion in the linear coordinate.

A more useful set of functions can be generated if we consider the case of nonazimuthal symmetry. It is straightforward, but relatively tedious to show that the general solution to Eq. (F.12) is given by

$$P_{lm} = (1-x^2)^{m/2}\frac{d^{(m)}}{dx^{(m)}}P_l(x). \tag{F.20}$$

This is called the *associated Legendre function*, and has the property [analogous to Eq. (F.18)] that

$$\int dx\, P_l^m(x)P_{l'}^m(x) = \frac{(l+m)!}{(l-m)!}\frac{2}{2l+1}\delta_{ll'} = c_l^m\delta_{ll'}. \tag{F.21}$$

The solution to the Laplace equation must then be of the form

$$\Phi = R(r)c_l^m P_l^m(x)e^{im\phi}.$$

The angular part of this function, containing the dependence of the solution on the angles θ and ϕ, is extremely important, and is given the name of *spherical harmonic*. It is written

$$Y_{lm}(\theta,\phi) = \left[\frac{2l+1}{4\pi}\frac{(l-m)!}{(l+m)!}\right]^{1/2}P_l^m(x)e^{im\phi}, \tag{F.22}$$

where we have inserted a factor of $1/\sqrt{2\pi}$ to normalize the function $e^{im\phi}$.

The spherical harmonics have the property that

$$\int Y_{lm}(\theta,\phi)Y_{l'm'}(\theta,\phi)\,d(\cos\theta)\,d\phi = \delta_{ll'}\,\delta_{mm'}, \tag{F.23}$$

i.e. they are an orthonormal set of functions. Unlike the sines and cosines or the Legendre polynomials, however, they are the basis vectors in a space of functions of two variables, rather than one. The extension of the idea of Appendix D to this case should be obvious.

This means that, just as we could expand any function defined on the interval $0 \le x \le L$ in a Fourier series, we can expand any function defined on the interval $0 \le \phi \le 2\pi$ $0 \le \theta \le \pi$ in a series involving spherical harmonics. Such a series would take the form

$$f(\theta, \phi) = \sum_{l,m} a_{lm} Y_{lm}(\theta, \phi), \tag{F.24}$$

where

$$a_{lm} = \int f(\theta', \phi') Y_{lm}(\theta', \phi') \, d(\cos \theta') \, d\phi. \tag{F.25}$$

Such expansions are extremely important in problems dealing with spherical geometries, such as problems relating to motions on the surface of the earth or deformations of a nucleus.

There remains a third set of coordinates which we used in the text, and this was the cylindrical. Laplace's equation in cylindrical coordinates is

$$\frac{\partial^2 \Phi}{\partial r^2} + \frac{1}{r} \frac{\partial \Phi}{\partial r} + \frac{1}{r^2} \frac{\partial^2 \Phi}{\partial \phi^2} + \frac{\partial^2 \Phi}{\partial z^2} = 0. \tag{F.26}$$

If we proceed as in Eq. (F.6), and assume that the solution is separable, so that

$$\Phi = R(r)Q(\phi)Z(z),$$

then tracing the steps from Eq. (F.8) to Eq. (F.12) yields

$$Z(z) = e^{\pm kz}$$

and

$$Q(\phi) = e^{\pm in\phi},$$

while the function $R(r)$ is determined by the equation

$$\frac{d^2 R}{dx^2} + \frac{1}{x} \frac{dR}{dx} + \left(1 - \frac{n^2}{x^2}\right) R = 0, \tag{F.27}$$

where we have set $x = kr$. This is called Bessel's equation, and the solutions to it are called *Bessel functions*.

We can determine the form of the Bessel functions just as we determined the Legendre polynomials. Assuming a power series solution of the form

$$R(x) = x^\alpha \sum_{j=0}^\infty a_j x^j, \tag{F.28}$$

we find, in analogy to Eq. (F.16), that

$$a_{2j} = \frac{-1}{4j(j + \alpha)}a_{2j-2}$$

and

$$\alpha = n,$$

so that the Bessel function is simply a power series in r, given by

$$J_n(x) = \left(\frac{x}{2}\right)^n \sum_{j=0}^{\infty} \frac{(-1)^j}{j!\,\Gamma(j + n + 1)}\left(\frac{x}{2}\right)^{2j}, \tag{F.29}$$

where we follow the usual convention and set $a_0 = [2^n\,\Gamma(n + 1)]$. An important difference in this case is that the series does not terminate, but includes all values of n.

The function $J_n(x)$ is called the *Bessel function of order n*. It has the general property that the function oscillates around zero, as shown schematically in Fig. F.1. We can denote by $x_{\nu n}$ the value of x for which the Bessel function of order n becomes zero for the νth time. It then follows (see the texts in the bibliography) that the Bessel function must have the orthogonality relation

$$\int_0^a x J_n\left(x_{\nu' n}\frac{x}{a}\right)J_n\left(x_{\nu n}\frac{x}{a}\right)dx = \frac{a^2}{2}[J_{n+1}(x_{\nu n})]^2\delta_{\nu\nu'},$$

which is an orthogonality condition similar to Eq. (F.18) for Legendre polynomials.

Obviously, the Bessel functions can be expected to play an important role in problems involving cylindrical symmetry, such as flow of the blood in an artery. The reader should be able to construct for himself the *Bessel series*, which is the analogue to the Fourier series, in terms of which functions whose argument runs from $0 \leqslant r \leqslant a$ can be expanded.

There are, of course, many more sets of orthogonal polynomials which are of use in specialized problems. In this text, only the spherical

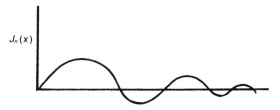

Fig. F.1.   A typical Bessel function.

harmonics and the Bessel functions appear, and the student will be able to handle almost all material which he encounters if he has a grasp of these basic functions and the ideas and concepts which underlie their use.

One final point should be made. We have mentioned that these new functions have a property of orthogonality, but we have nowhere shown that they form a complete set of basis functions. In fact, this is shown in most texts on differential equations, and need not disturb the reader unduly.

Index

Acoustic wave
  in a fluid, 85–86
  in a solid, 221–222
Arms control, 240, 244
Arterial walls
  composition, 250
  response to pressure, 252ff.
Arteriosclerosis, 263
Artery, 249: see also blood flow

Bénard cell: see convection cell
Bernoulli equation, 57, 65
Bessel
  equation, 258, 267, 298
  function, 298–299
Biharmonic equation, 232
Blood
  cells, 250
  composition, 249
  flow, 249ff.
    arterial, 256ff.
    Reynolds number for, 152
Bonneville, 131, 136
Borda's mouthpiece, 66
Bossinesq approximation, 158, 172
Boundary layer, 145
Boundary-layer separation, 151
Boundary-value problem, 294
Breakaway, 241

Buckling, 200
Bulk modulus, 219

Cantilever, 204, 207
Capillary
  in blood circulation, 249
  jet, 265ff.
  wave, 81–83
Circulation, 66
Circulatory system, 248ff.
Collagen, 250–251
Complex potential, 66
Continental drift, 179, 195, 240
Continuity
  equation, 5–8
  for plane surface, 71
  for spherical surface, 96, 103
Convection cells, 168–170
  in the atmosphere, 171, 172
  and continental drift, 179
Convective derivative, 2–3
Core (of the earth), 240
Coriolis force, 93
Crust (of the earth), 239

Diffusivity, coefficient of, 157
Disturbing potential, 89
  at equator, 90
  general form, 97

Doldrums, 172
Drag, 138, 152

Earth
  as a fluid, 27
  free oscillations, 111
  viscosity of, 130–136
Elastic constants, 189, 217–220
Elastic solid, 188
Elastin, 250–251
Entropy flux density, 13
Entry problem, 264
Equation of state, 9, 17
  polytropic, 12
Equilibrium
  neutral, 41
  stable, 40, 58
  systems far from, 118–119
  types of, 39
  unstable, 41, 58
Euler equation, 4–5, 123
  in a galaxy, 30
  for potential flow, 57
  for rotation, 18–19, 93
Euler theory of struts, 199
Expansion, coefficient of, 158

Fenno-Scandian uplift, 131, 136, 203, 204
First law of thermodynamics, 156
Fission
  induced, 119
  of a nucleus, 117–119
  spontaneous, 116
Fissionability parameter, 116–118
Fluid
  classical, 2, 122, 164, 184
  incompressible, 7
Fourier series, 288, 291, 293

Galileo, 138
Grashof number, 180
Green's theorem, 283

Hadley cell, 171, 177
Heat equation, 156
Heat transfer equation, 181
Hilbert space, 287
Hooke's law, 188, 190, 209, 217, 218, 230,
    250, 259
Horse latitudes, 172
Huygens principle, 235
Hydraulics, 144

Ideal gas law, 9
Incompressible fluid, 7, 84
  earth as an incompressible fluid, 27
Inversion, thermal, 164
Irrotational flow, 56

Jacobi ellipsoids, 27

Kronecker delta, 278

Lacolith, 192–195, 205
Lamé coefficients, 217–220, 230
Laplace
  equation, 75, 292, 294, 298
  equation for potential flow, 56
  theory of the tides, 90, 102
Legendre
  equation, 295
  function, 296–297
Linearization, 61, 63, 73
Liouville Theorem, 65
Loading
  critical, 200
  of a solid, 186
Long waves, 68, 74
Longitudinal wave in a solid, 221
Love waves, 227–229, 261
Lubrication, theory of, 153

Mach number, 153
Maclaurin ellipsoids, 22, 28
  stability of, 43–47
Mantle, 239
Membrane tension, 254
Micron, 250
Mountain chain, 195–199
Mohorovičič discontinuity, 240

Navier–Stokes equation, 127
Neutral filament, 190
Newton's second law of motion, 2, 5, 123,
    220, 252
Newtonian solid, 217, 230, 250

Normal modes of oscillation
    for the earth, 110
    for oceans, 104

Orr–Sommerfeld equation, 273

P wave, 223, 235, 238
    in nuclear explosions, 244
    reflection, 245–256
Pisa, leaning tower, 138
Poisieulle
    flow, 127, 152
    formula, 130, 271
Poisson
    equation, 9, 13
    ratio, 189, 192, 218
    relation, 226
Potential flow, 56
Prandtl, 144
    equations, 148
    number, 180

Rayleigh criterion
    for convection, 167, 180
    for jet stability, 270, 272
Rayleigh, Lord, 167, 180, 226
Rayleigh wave, 226–227
    in a nuclear explosion, 244
Reynolds number, 142
    for blood flow, 152, 272
Rheology, 188
Ripple, 83
Roche's limit, 47–48

S wave, 223, 235, 238
    in nuclear explosion, 244
    reflection, 245–246
SH wave, 225, 227
SV wave, 225, 227
    reflection of SH and SV waves,
        245–256
Second sound, 182
Seiche, 84
Seismic radius, 241
Seismic ray parameter, 238
Separation of variables, 292
Shear force
    in a fluid, 123

    at a fluid surface, 135
    in solids, 212, 214, 219
Shear modulus, 219
Shear wave
    in a solid, 222
    horizontal, 225
Similarity, law of, 144, 152, 180
Slip, 129
Smog, 164
Snell's law, 236, 246
Solar wind, 14
Sound wave in a fluid, 85–86
Specific heat, 156
Spherical harmonics, 297
Stoke's
    first problem, 139
    formula, 138
    second problem, 140
Strain tensor, 210–212, 214, 217
Stream function, 66, 67, 149, 153, 273
Streamline, 65
Stress function, 232
Stress tensor
    Maxwell, 215
    for a solid, 212–216, 217, 218, 230, 252
    for viscosity, 124, 126
Strong interactions, 112
Struts, 200
Summation convention, 277
Superfluid, 182
Surface tension, 79, 254

Tamped explosion, 241
Tangential instability, 59
Tensor
    Cartesian, 3, 276ff.
    Maxwell stress, 215
    momentum flux, 8, 13
    strain, 210–212
    stress, 212–216, 230
    viscous, 124
Thermal conductivity
    coefficient of, 156
Tides, 88
    diurnal, 98, 101, 103
    equatorial, 89
    inverted, 92, 101
    monthly, 98

Tides (*continued*)
  planetary, 102
  semi-diurnal, 91–92, 98, 101
  solar, 92, 102
Torsional rigidity, 231
Tuning fork, 233

Urethra, 265, 270
Urinary drop spectrometer, 265ff.
Urinary system, 264ff.

Velocity field, 49
Velocity potential, 56
  for capillary jet, 267
  for surface waves, 75

Viscoelastic solid, 250, 256
Viscosity, 122ff.
  coefficients of, 126
  energy, 137–138
  kinematic coefficient of, 127
Vorticity transport equation, 138

Waves in solids
  body, 220–223
  surface, 223–227
  thin sheets, 232, 233
Wronskian determinant, 163

Young's modulus, 188, 218, 230

Zonal heating, 172